Advective Transport Observations with MODPATH-OBS–Documentation of the MODPATH Observation Process

By R.T. Hanson, L.K. Kauffman, M.C. Hill, J.E. Dickinson, and S.W. Mehl

Chapter 42 of
Section A, Groundwater
Book 6, Modeling Techniques

Prepared in cooperation with the U.S. Department of Energy

Techniques and Methods 6–A42

U.S. Department of the Interior

U.S. Geological Survey

U.S. Department of the Interior
SALLY JEWELL, Secretary

U.S. Geological Survey
Suzette M. Kimball, Acting Director

U.S. Geological Survey, Reston, Virginia: 2013

For more information on the USGS—the Federal source for science about the Earth, its natural and living resources, natural hazards, and the environment, visit http://www.usgs.gov or call 1–888–ASK–USGS.

For an overview of USGS information products, including maps, imagery, and publications, visit http://www.usgs.gov/pubprod

To order this and other USGS information products, visit http://store.usgs.gov

Suggested citation:
Hanson, R.T., Kauffman, L.K., Hill, M.C., Dickinson, J.E., and Mehl, S.W., 2013, Advective transport observations with MODPATH-OBS—Documentation of the MODPATH observation process: U.S. Geological Survey Techniques and Methods book 6—chap. A42, 96 p., http: pubs.usgs.gov/tm/06/a42.

Preface

The MODPATH-OBS computer program described in this report is designed to calculate simulated equivalents for observations (or predictions) related to groundwater transport that can be represented in a meaningful way using particle tracking. Generally this means that the observations relate to the movement of the center of a discrete plume, movement of position along a plume front selected to correct for the effects of dispersion and reactions not simulated by particle tracking, geochemical data that identify likely recharge areas, and other hydrologic features.

MODPATH-OBS uses the particle tracking capabilities of MODPATH by acting as a postprocessor. MODPATH is a postprocessor for MODFLOW, so that the sequence of model runs generally required is MODFLOW, MODPATH, and MODPATH-OBS.

The versions of MODFLOW and MODPATH that support MODPATH-OBS as documented in this report are MODFLOW-2000/2005 and MODFLOW LGR (Mehl and Hill, 2005, 2007) and MODPATH (Pollock, 1994, 2012) or MODPATH-LGR (Dickinson and others, 2011). MODFLOW-LGR is derived from MODFLOW-2005 (Harbaugh, 2005) and supports local grid refinement. MODFLOW-LGR simulations may include local grid refinement or may use a single grid. When a single grid is used, MODFLOW-LGR performs identically to the version of MODFLOW-2005 cited on the MODFLOW-LGR web page. MODPATH-LGR and MODPATH-OBS can use nearly all of the capabilities of MODFLOW-LGR. For example, simulations can be steady-state, transient, or a combination; model layers may be confined or convertible from confined to unconfined; all of the head-dependent boundary packages are supported and features simulated using the Streamflow-Routing (SFR) Package can be routed across grid boundaries. Limitations are discussed in this report.

MODPATH-LGR (Dickinson and others, 2011) is derived from MODPATH (Pollock, 1989, 1994, 2012) and supports the tracking of particles through locally refined grids. For a single grid and no observations, MODPATH-LGR performs similarly to the version of MODPATH cited on the MODPATH-LGR web page. Differences occur mostly in that the output files now include information about the grid through which the particle travels.

MODPATH-OBS is intended to replace the capabilities of the Advective-Transport Observation (ADV2) Package of MODFLOW (Anderman and Hill, 2001), which was limited to steady-state flow fields. MODPATH-OBS provides advances such as the support of transient flow fields, added observation types, and local grid refinement, as described in this report.

MODPATH-OBS is primarily intended for use with separate programs that conduct sensitivity analysis, data needs assessment, parameter estimation, and uncertainty analysis, such as UCODE-2005 (Poeter and others, 2005) and PEST (Doherty, 2007). Though the program name specifically refers to observations, the quantities calculated can also be model predictions.

The documentation presented here describes the methods and their utility, and input and output files. An example is used to demonstrate MODFLOW-LGR, MODPATH-LGR, and MODPATH-OBS.

The code for this model is available for download over the Internet from a U.S. Geological Survey software repository. The repository is accessible from the U.S. Geological Survey Water Resources Information web page at: http://water.usgs.gov/software/ground_water.html.

The performance of the MODPATH-OBS program has been tested in a variety of applications. Future applications, however, might reveal errors that were not detected in the test simulations. Users are requested to notify the U.S. Geological Survey of any errors found in this document or the computer program using the email address available on the website. Updates might occasionally be made to both this document and to the MODPATH-OBS program, and users are encouraged to check the website.

Contents

Contents

Figures

Tables

Conversions, Abbreviations, and Acronyms

Inch/Pound to SI

Multiply	By	To obtain
Length		
foot (ft)	0.3048	meter (m)
Area		
acre	4,047	square meter (m^2)
square foot (ft^2)	0.09290	square meter (m^2)
Volume		
gallon (gal)	0.003785	cubic meter (m^3)
cubic foot (ft^3)	0.02832	cubic meter (m^3)
Flow rate		
acre-foot per day (acre-ft/d)	0.01427	cubic meter per second (m^3/s)
cubic foot per second (ft^3/s)	0.02832	cubic meter per second (m^3/s)

SI to Inch/Pound

Multiply	By	To obtain
Length		
meter (m)	3.281	foot (ft)
Area		
square meter (m^2)	0.0002471	acre
square meter (m^2)	10.76	square foot (ft^2)
Volume		
cubic meter (m^3)	264.2	gallon (gal)
cubic meter (m^3)	35.31	cubic foot (ft^3)
Flow rate		
cubic meter per second (m^3/s)	70.07	acre-foot per day (acre-ft/d)
meter per second (m/s)	3.281	foot per second (ft/s)

Abbreviations and acronyms are defined as follows:

cfc	chlorofluorocarbon
MF2K5	MODFLOW-2005
MODFLOW-LGR	MODFLOW-2005 with Local Grid Refinement
MODPATH	MODPATH ver 6 without Local Grid Refinement
MODPATH-LGR	MODPATH ver 6 with Local Grid Refinement
MODPATH-OBS	MODPATH Observations with Local Grid Refinement
MF-FMP	MODFLOW-2005 version 1.6 with the Farm Process version 2
pce	perchloroethylene
PEST	Universal parameter estimation code
SF_6	Sulfur hexaflouride
SAMM	Southern Amargosa Embedded Model
USGS	U.S. Geological Survey
UCODE-2005	Universal parameter estimation code

Advective Transport Observations with MODPATH-OBS–Documentation of the MODPATH Observation Process

By R.T. Hanson, L.K. Kauffman, M.C. Hill, J.E. Dickinson, and S.W. Mehl

Abstract

The MODPATH-OBS computer program described in this report is designed to calculate simulated equivalents for observations related to advective groundwater transport that can be represented in a quantitative way by using simulated particle-tracking data. The simulated equivalents supported by MODPATH-OBS are (1) distance from a source location at a defined time, or proximity to an observed location; (2) time of travel from an initial location to defined locations, areas, or volumes of the simulated system; (3) concentrations used to simulate groundwater age; and (4) percentages of water derived from contributing source areas. Although particle tracking only simulates the advective component of conservative transport, effects of non-conservative processes such as retardation can be approximated through manipulation of the effective-porosity value used to calculate velocity based on the properties of selected conservative tracers. This program can also account for simple decay or production, but it cannot account for diffusion. Dispersion can be represented through direct simulation of subsurface heterogeneity and the use of many particles.

MODPATH-OBS acts as a postprocessor to MODPATH, so that the sequence of model runs generally required is MODFLOW, MODPATH, and MODPATH-OBS. The version of MODFLOW and MODPATH that support the version of MODPATH-OBS presented in this report are MODFLOW2000/2005 or MODFLOW-LGR, and MODPATH or MODPATH-LGR. MODFLOW-LGR is derived from MODFLOW-2005, MODPATH 5, and MODPATH 6 and supports local grid refinement. MODPATH-LGR is derived from MODPATH 5. It supports the forward and backward tracking of particles through locally refined grids and provides the output needed for MODPATH-OBS. MODPATH-LGR and MODPATH-OBS simulations can use nearly all of the capabilities of MODFLOW-2005 and MODFLOW-LGR; for example, simulations may be steady-state, transient, or a combination. Though the program name MODPATH-OBS specifically refers to observations, the program also can be used to calculate model prediction of observations.

MODPATH-OBS is primarily intended for use with separate programs that conduct sensitivity analysis, data needs assessment, parameter estimation, and uncertainty analysis, such as UCODE_2005, and PEST.

In many circumstances, refined grids in selected parts of a model are important to simulated hydraulics, detailed inflows and outflows, or other system characteristics. MODFLOW-LGR and MODPATH-LGR support accurate local grid refinement in which both mass (flows) and energy (head) are conserved across the local grid boundary. MODPATH-OBS is designed to take advantage of these capabilities. For example, particles tracked between a pumping well and a nearby stream, which are simulated poorly if a river and well are located in a single large grid cell, can be simulated with improved accuracy using a locally refined grid in MODFLOW-LGR, MODPATH-LGR, and MODPATH-OBS. The locally-refined-grid approach can provide more accurate simulated equivalents to observed transport between the well and the river.

The documentation presented here includes a brief discussion of previous work, description of the methods, and detailed descriptions of the required input files and how the output files are typically used.

Introduction

Problem

Using concentration data directly as observations or locations in nonlinear regression techniques to develop models capable of simulating advection, dispersion, and reactions is complicated by large computational demands and numerical dispersion (Zheng and Bennett, 1995; Barlebo and others, 1998; Mehl and Hill, 2000, 2001). Simulation of predictions faces similar problems. The difficulties tend to become greater as the simulations become larger. Therefore, methods to extract fundamental groundwater flow system information from concentration data without resorting to the use of the advection-dispersion equation are of interest (Sanford, 2011).

Particle tracking, which generally represents the advective component of transport, provides such an opportunity for a simpler alternative when used with observations derived from measured concentrations (Anderman and others, 1996). In some cases, investigations use particle tracking as a preliminary step in groundwater investigations and then continue on to use the more complicated transport capabilities; in other cases, the analysis using particle tracking can directly serve the purpose of the modeling effort.

Efforts to compare simulated groundwater ages from particle tracking with measured age tracer data is made difficult because mixing during groundwater flow or well-bore flow may produce field samples with wide age distributions. Mixing may occur within a well bore or be caused by preferential flow, dual porosity processes, subsurface characteristics that cause divergent or convergent flow (such as karst features or heterogeneity within porous media), or flows to and from surrounding water bodies such as lakes and rivers.

Large scale, regional models typically have physically based boundary conditions, but grids generally are too coarse to accurately represent local boundary conditions and related drawdown near pumping wells. Representing local stresses such as pumping, distribution of inflow to a well bore in multi-aquifer wells, flows from springs and rivers, or aquifer tests in complex hydrogeologic settings can be difficult because coarse grid discretization can cause numerical dispersion and distortion of advective-travel paths. Local-scale models can adequately discretize a system to represent local groundwater conditions, but these models usually have tenuous outer boundary conditions. These issues affect simulation of advective transport as well as groundwater flow. Analytical methods often are not appropriate due to non-ideal conditions, such as heterogeneity and (or) anisotropy.

Locally-refined grids provide an opportunity to represent local conditions adequately while maintaining a rigorous numerical connection to the regional system. Regional and local models are linked in a way that maintains continuity of hydraulic head and flow across the shared boundary. In locally-refined models, parameters can by defined throughout the simulated system. Parameters and observations in the regional model can be important to the local model, and vice versa. For example, one parameter might be defined to represent the same rock type in both the regional and local models, and simulated-equivalent observations may be affected by system dynamics simulated in more than one model.

Advective transport observations may include travel time, proximity or distance to actual location, concentration, and fractions of source waters. These observations frequently are affected by processes that occur at local and regional scales. Estimating parameters at different scales using locally-refined models is advantageous, but it seldom is practiced. The incorporation of transport information from particle tracking can provide important new information for developing such simulations. MODPATH-OBS provides a way of addressing these problems through the integration of advective-transport observations from regional grids linked to locally-refined grids.

For some situations, particle tracking can be used to design monitoring and sampling systems. For example, if the probable magnitude of travel times or mixtures of water from source areas are of interest, MODPATH-OBS can be used to identify observations likely to improve the accuracy and reduce the uncertainty of simulated results.

Purpose and Scope

The purpose of this report is to document MODPATH-OBS and its use with output from MODFLOW/ MODPATH. The report includes sections describing the simulated quantities that can be produced by MODPATH-OBS and how these simulated quantities can be used in model development, including sensitivity analysis, calibration, prediction, uncertainty, and data-needs assessment. The four types of observations (simulated equivalents for observations related to advective groundwater transport) that can be meaningfully represented by using MODPATH-OBS are as follows:

1. Proximity—The proximity of one or more transported particles to an observed location, or the predicted travel distance. The proximity or distance in one, two, or three dimensions can be defined, or a Euclidean distance can be used.

2. Time of travel—The time it takes groundwater to flow from a user-specified source location to a defined location or boundary. The boundary may be defined as an area or volume.

3. Concentration—Commonly used to represent water age when the source history can be reasonably estimated, and also able to represent other types of chemical or isotopic observations. In MODPATH-OBS, each particle can be associated with a starting concentration. The concentration can be affected by one degradation rate along the travel path, and values from multiple particles can be averaged at the destination location. Volume weighting of concentrations is also supported.

4. Source-water type—Used to simulate sources of water contributing to a water sample. Sources might be, for example, recharge from certain areas, leakage from surface water, or regional inflow. In MODPATH-OBS, each particle is associated with a volume from the relevant source location. The volumes for each particle associated with a source are summed and then divided by the total volume of a sample, to express the percent of a given source-water type.

All observation types can involve transport from one point to another, as might occur when representing movement of a plume center-of-mass or plume front. Transport starting and ending locations generally can also be defined using areas or volumes; restrictions are described in this document. All observation types can involve forward or backward tracking of particles.

The models used with MODPATH-OBS can have a single grid or can have a locally-refined grid (or grids) embedded within a regional grid. Locally-refined grids are often important to obtaining accurate flow fields and associated particle tracking results. The example included in this report demonstrates this capability.

PEST, UCODE-2005, and other parameter estimation programs can be used with results from MODPATH-OBS to do sensitivity analysis, inverse modeling with locally refined grids, predictions, and uncertainty of the predictions. This report provides an example of sensitivity analysis and parameter estimation using nonlinear regression conducted with MODPATH-OBS and UCODE_2005. For predictions, each observation type can be used to simulate model predictions, for which no observed equivalent of field-based measurements exists.

The ability to include computationally efficient particle-tracking observations in the development of locally refined models requires a systematic method of generating simulated values to compare to observations, as is provided by MODPATH-OBS. Using sensitivity analysis methods, MODPATH-OBS allows users to quantify the value of transport observations relative to improving the conceptual consistency and accuracy of flow and transport models. The general process could be used as part of model verification as mandated by the Underground Tank Assessment (UGTA) strategy (Dixon and Peterson, 2003). Observations provide additional simulation features and constraints for parameter estimation of hydraulic properties.

Previous Work

Particle-tracking has been used to produce simulated equivalents to observations for model calibration, sensitivity analysis, uncertainty evaluation, prediction, and data-needs assessment in many studies. Selected works are listed in table 1. In the absence of numerical flow models, lumped parameter models such as TRACERMODEL1 have been used to estimate age distributions of groundwater samples on the basis of age tracer data (Cook and Böhlke, 2000; IAEA (International Atomic Energy Agency), 2006). Campana and Simpson (1984) provided an early demonstration of using C^{14} age dates to estimate residence times and recharge rates with a simple discrete-state compartment mixing model.

Many studies have employed particle tracking in numerically simulated flow fields using codes such as MODPATH (Pollock (1989, 1994) and ADV2 (Anderman and Hill, 2001).

Zimmerman and others (1991) compared parameter estimation and sensitivity analysis techniques and their effect on uncertainty in groundwater flow model predictions by using MODPATH particle-tracking with the groundwater flow model of Avra Valley, Arizona (Hanson and others, 1990; Hanson, 1996). This early study indicated that the cumulative distribution function (CDF) of particle travel times was greatly influenced by the forms of boundary conditions and the estimation of hydraulic properties. The CDF was also greatly influenced if simulation results were censored to include only realizations that were physically possible based on the conceptual constraints of the real flow system, such as censoring the realizations by omitting those with groundwater levels above the land surface. The distribution of flow paths throughout the flow system for the ensemble of transmissivity field realizations was also determined to be an important aspect of the observations from particle tracking that could provide an additional constraint for parameter estimation.

Many investigators use particle tracking to qualitatively assess the relationship between recharge components and geochemical data collected at wells or streams down gradient from these sources of groundwater inflow. For example, Reichard and others (2004) compared particle tracking from artificial recharge areas into the Los Angeles Basin, and Izbicki and others (2004) compared separate particle-tracking runs for various natural recharge sources with groundwater age data. These studies qualitatively compared the results of particle tracking with stable-isotopic data and ^{14}C age-dates but did not use these data to make quantitative comparisons or use the data as observations for model calibration. Hunt and others (2005) used deuterium and oxygen isotopes to look at the potential travel times of poor water-quality streamflow infiltration and compared these with particle tracking estimates of travel times of simulated infiltration. Werner and others (2006) used ^{225}Rn isotopic data for qualitative comparison of stream-aquifer interactions for a regional flow model of a tropical watershed in Australia.

Sanford and others (2003, 2004) used observations of water age to calibrate a model of the Middle Rio Grande Basin groundwater flow model. The water age observations were based on a suite of geochemical tracers (percent modern carbon, and deuterium and oxygen stable isotopes) that were sampled and analyzed regionally to provide field-based observations (Plummer and others, 2004). Problem-specific computer programs were used to generate simulated values to compare with mixtures of source-area waters derived from stable isotopes. The design of the MODPATH-OBS is based, in part, on this pioneering work.

Table 1. Selected works that compare field data with particle tracking results.

Field conditions	Observation[1]	Observation type in this work	Selected references[2]	
Plume from tracer test or otherwise, originates from known location	Time and path of travel	Time of travel proximity	[3]Sykes and Thomson, 1988 [3]Anderman+, 1996 [3]Anderman, Hill, 2001 [3]Tiedeman+, 2003, 2004	
Environmental tracer natural or anthropogenic, originating from known location or area	Age dating	Time of travel, concentration	Campana and Simpson, 1984 Phillips+, 1989 Cook and Böhlke, 2000 IAEA, 2006 Zhu, 2000 Plummer+, 2004, 2013	[3]Sanford+, 2003, 2004, 2011, 2013 Werner+, 2006 Bethke and Johnson, 2008 [3]Kauffman and Crandall, 2008 Torgeson+, 2013 Engdahl+, 2013 Ginn, 1999, 2000a, 2000b, 2007 Ginn+, 2009
Age of water arriving at a well over time	Percent of time one or more ages are exceeded	Concentrations using the exceedance curve option	Hanson+, 1990 Hanson, 1996 [3]Zimmerman+, 1991	
Groundwater velocity from in situ temperature probe	Local velocity	Time of travel	Ballard, 1996	
Groundwater velocity from temperature perturbation	Local velocity, usually under surface-water body	Time of travel	Constantz+, 2003	
Well-bore flow measurements	Local velocity in a well bore	Time of travel	Newhouse and Hanson 2000, 2002 Clark +, 2008	
Continuous source plume		Proximity time of travel	Domenico and Schwartz, 1990 p. 362	
Leading edge of plume	Arrival time	Normally ill advised[4]	Cook, Böhlke, 2000	
Source of water to a well. Capture by pumping.	Stable isotopes like deuterium, oxygen. Possibly augmented by age dating	Source water	Muir and Coplen, 1981 Franke+, 1998 [3]Starn, 2000 Hanson+, 2002 Newhouse+, 2004 Starn, Stone, 2004 Reichard+, 2004	Izbicki+, 2004 Plummer+, 2004 Sanford+, 2003, 2004 Hunt+, 2005 Clark+, 2008 Burow+, 2008 Leake+, 2008, 2010
Depth dependent sampling at a supply well			Izbicki+, 1999 Hanson+, 2002 Clark+, 2008	

[1]Potential problems include no simulated mixing along the flow path and insufficient sampling to define a desired quantity, such as a plume center or a plume front. See the text for a discussion of likely mechanisms.

[2]+, and others.

[3]Advective-travel observations using in model analysis, including sensitivity analysis, parameter estimation, and (or) uncertainty evaluation. This approach may include predictions.

[4]Need detailed subsurface, many particles. Possibly use piston exponential model.

Clark and others (2008) analyzed the potential vulnerability to contamination of public supply wells in a study area in Nebraska and assessed the contributing recharge areas to public supply wells through particle tracking and comparisons with measured age tracer and chemical data. The particle tracking results were consistent with field data in indicating that wells screened in multiple aquifers were the primary source of poor-quality young water reaching public supply wells screened only in a confined aquifer. In addition, particle-tracking simulations compared with age tracer data were used to estimate age distributions of water reaching supply wells and make long-term projections of the effects of denitrification of nitrate-nitrogen concentrations based on simulated groundwater flow and denitrification rates estimated from field data. Simulated and observed ages and percentages of young water were compared. This analysis provided a first-order comparison of the effects of denitrification rates under steady-state flow (Clark and others, 2008). This study also applied techniques for forward-tracking particles through multi-aquifer well bores that facilitated the internal mixing through the generation of intermediate particles. Tracking and splitting particles through well-bore flow is not implemented in this version of MODPATH-OBS.

Burow and others (2008) used particle pathline analysis to analyze the contributing area and age distribution of groundwater withdrawn from a public-supply well in Modesto, eastern San Joaquin Valley, California. The particle tracking results were compared to age tracer data in the public-supply well and in monitoring wells along a groundwater flowpath to the public-supply well.

Age dates have been used from measured tritium and sulfur hexafluoride (SF6) concentrations in the calibration of a flow and advective-transport model near public-supply wells (Lindgren, 2011; Crandall and others, 2008; Kauffman and Crandall, 2008; Burow and others, 2008). These studies used the input concentrations history for comparison with various observation points in the flow system to derive groundwater ages as additional constraints on the calibration of flow-model hydraulic properties. Adding the age-date observations to the flow and head observations improved model fit in the calibration process.

MODFLOW-OBS allows the kinds of analyses described above to be conducted using grids in which local refinement is introduced to improve the simulation of, for examples, geologic structure, hydraulics around wells and rivers. Examples applications of using refined grids to simulate local effects of pumping wells include Graham and Smart (1980), von Rosenberg (1982), Székely (1998), Mehl and others (2006), and Dickinson and others (2007). The ability to route streamflow across local grid boundaries can be simulated using MODFLOW-LGR (Mehl, 2008).Mehl and Hill (2003) and Keating and others, (2003) used sensitivity analysis and inverse modeling with locally refined grids.

The collection, analysis, and use of environmental data for comparison purposes are beyond the scope of this user's manual for MODPATH-OBS. However, numerous studies have been completed that demonstrate how these types of data could be extended to a more quantitative level of comparison with respect to simulation of hydrologic flow systems and related model calibration, sensitivity, and uncertainty analysis. The reader is referred to the summaries of regional water quality analysis such as Alley (1993) and to compilation of environmental isotopic tracers such as Clark and Fritz (1997) or Cook and Herczeg (2000) for information required to prepare and synthesize field data into estimates that become useful comparisons. The reader is also referred to specific studies that applied field data such as the application of stable isotopes to estimate percentages of source water for the Santa Clara Valley, California (Muir and Coplen, 1981; Hanson and others, 2002; Newhouse and others, 2004) or the middle Rio Grande Basin, New Mexico (Sanford and others, 2004); or the application of carbon-14 data to regional flow systems modeling to estimate hydraulic properties and recharge rates such as the San Juan Basin, N. M. (Phillips and others, 1989), the Black Mesa, Ariz. (Zhu, 2000), and the Middle Rio Grande Basin, N. M. (Sanford and others, 2004).

Overview

This report discusses how MODPATH-OBS simulated equivalents of observations can be constructed and compared with field values, describes the calculations performed by MODPATH-OBS and their limitations, and presents a hypothetical example that is used to demonstrate MODPATH-OBS. Also this report provides detailed input instructions for MODPATH-OBS and describes selected input to a hypothetical example problem for steady-state and transient-state simulations.

The hypothetical problem is used to demonstrate the method and all of the observation features in MODPATH-OBS. A regional model and two embedded child models are represented in the hypothetical problem. This model is used to simulate system changes in groundwater conditions resulting from local pumping and regional recharge using a locally refined model simulated using MODFLOW-LGR (Mehl and Hill, 2005) and MODPATH-LGR (Dickinson and others, 2011). Files for steady-state and transient flow conditions are distributed; results from the transient model are presented in this report. Parameter estimation is used with particle-tracking observations to assess the sensitivity and uncertainty of hydraulic conductivities and porosities of selected zones within the parent and child models by using UCODE-2005 (Poeter and others, 2005).

MODPATH-OBS can be used with a single grid or with locally refined grids. MODPATH-OBS can be used with standard MODFLOW (Harbaugh, 2005) and MODPATH (Pollock, 2005, 2012) for single model-grid applications, but MODPATH-OBS depends on output from MODFLOW-LGR (Mehl and Hill, 2005) and MODPATH-LGR (Dickinson and others, 2011) for multiple grids that employ the methods developed by Mehl and Hill (2005, 2007).

Relating Simulated Advective Transport to Field Conditions

Recent articles suggest there are general difficulties that still exist in simulating subsurface transport (Clement, 2010; Konikow, 2011), including numerical issues, unknown field conditions, and sparse measurements. Advective transport is the simplest component of transport to calculate, but is effected by these difficult technical issues.

The utility of particle tracking in groundwater model development and prediction has been questioned because of the difficulties discussed in this section (for example, see Bethke and Johnson, 2008). While the technical challenges of particle tracking are real, it seems apparent that there are many circumstances in which advective-transport observations, as well as predictions, can be used beneficially in model development (Sanford, 2011). Particle tracking supports at least a rough quantitative analysis of available information on flow direction, distribution, and rate.

Particle tracking has at least two primary limitations.

1. It is unclear how to accommodate the effects of dispersion, retardation, decay, production, chemical reactions, a possibly transient flow field, unrepresented heterogeneity, and other factors common to field conditions that cause subsurface transport to differ from the plug flow represented by advective transport. In many instances, these problems are compounded by the absence of a reliable history of source release times, strengths, amounts, or locations (which are also a potential limitation to modeling solute transport with dispersion). As such, even quantified comparisons of particle-tracking results that are developed with MODPATH-OBS may only represent an upper or lower bound to potential transport within a complex groundwater flow field. Some examples include the following:

   a. Models that lack the vertical resolution (model layers) needed to explicitly simulate more transmissive zones that typically occur in fluvial and marine sedimentary deposits can produce anomalous underestimates of porosity to match the relatively faster observed transport.

   b. If not accommodated by adjustment of field observations or the simulation, dispersion, retardation, and decay can cause underestimation of the actual rate of advective transport; similarly, production can cause overestimation. Dispersion can be addressed in the interpretation of concentration data to obtain measures of advective transport, as is discussed below. Retardation can be addressed by adjusting the porosity used in MODPATH (or

MODPATH-LGR). Decay or production can be addressed when using the concentration observation type available in MODPATH-OBS.

2. Abrupt lateral or vertical changes in hydraulic conductivity can result in particle tracks that change dramatically as hydraulic conductivity changes (LaVenue and others, 1989; Poeter and Gaylord, 1990). Similarly, abrupt watershed divides that occur in topographically driven flow fields with nested watersheds of different scale create abrupt changes in ages across the region. This leads generally to poor matches between observed and advectively simulated ages. These abrupt changes can violate smoothness requirements of gradient-based optimization methods, such as the modified Gauss-Newton method used in MODFLOW-2000 (Hill and others, 2000), UCODE_2005, and PEST. An example of this problem is presented in the Common Calibration Problems section of this report. Severe situations can diminish the utility of sensitivity analyses and optimization methods such as those described by, for example, Hill and Tiedeman (2007).

A simple analysis can provide some insight into problems related to dispersion. Here the problem of locating the advective-front of a plume is considered (fig. 1A). This situation is most directly related to the proximity and time-of-travel types of observations and predictions discussed below, but consequences can be substantial for all types of observations and predictions considered in this report.

For a continuous source in a one-dimensional flow field with homogeneous hydraulic conductivity in the presence of longitudinal dispersion only, the advective front is located along the plume centerline at the 50-percent-concentration contour (defined as the concentration halfway between the source concentration and the background concentration). In this situation, longitudinal dispersion causes the contaminant front to spread out along the length of the plume, but the advective front is always located at the farthest point from the source on the 50-percent concentration contour (Domenico and Schwartz, 1990, p. 362). By adding the single complexity of transverse dispersion, the plume front stays closer to the source for a given elapsed time of movement than does the advective front, and the 50-percent-concentration contour falls short of the advective front. This problem is illustrated in figure 1, which demonstrates the sensitivity of the location of the 50-percent-concentration contour to vertical transverse dispersivity. The presence of transverse dispersion results in the 50-percent concentration contour being located closer to the source than the advective front. The high sensitivity of the location of the 50-percent concentration contour to vertical transverse dispersivity in this problem is due to the small vertical dimension of the contaminant source relative to the horizontal dimension.

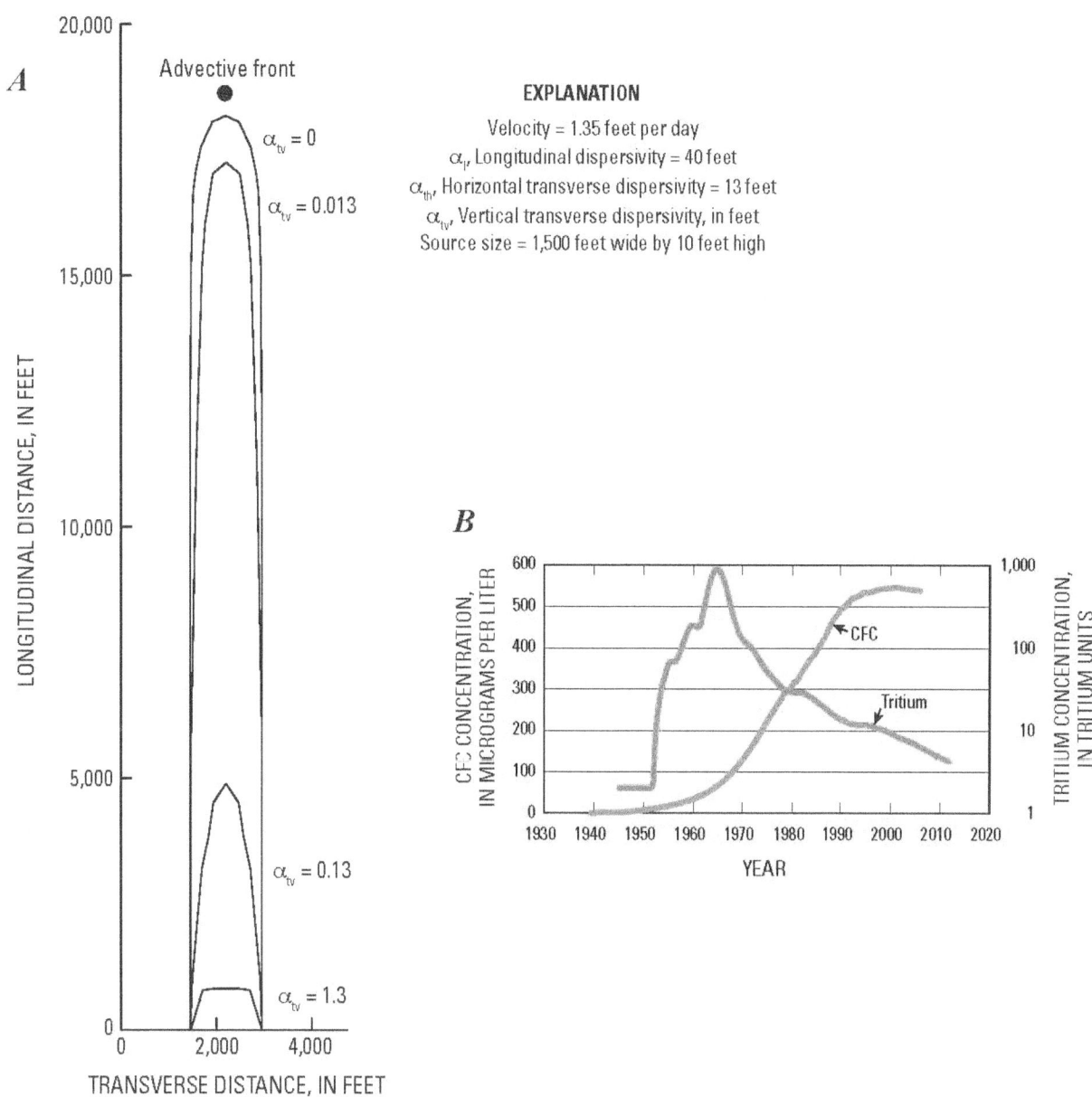

Figure 1. Two difficulties that can occur when relating field data to particle-tracking results. (*A*) Location of the 50-percent-concentration contour with varying vertical transverse dispersivity, calculated using Wexler's (1992) three-dimensional strip source analytical solution. The advective-front location, defined as the fluid velocity (here, 1.35 feet per day) times the time of travel (about 137 days), is located at the dot. (*B*) Historical concentration time series of chlorofluorocarbons (cfc's) and tritium showing a simple and complex source history, respectively (Jurgens and others, 2012).

In general, the location of the 50-percent-concentration contour will be more sensitive in situations where transverse dispersion is large in the direction in which the source has a small dimension. For example, the location of the 50-percent-concentration contour generated by a tall and narrow source is more sensitive to horizontal transverse dispersion than to vertical transverse dispersion, and the location of the 50-percent-concentration contour generated by a short and wide source is more sensitive to vertical transverse dispersion than to horizontal transverse dispersion.

While it is tempting to use particle tracking to represent arrival times, this is normally ill-advised. Regardless of how many particles are used, first arrivals generally can only be simulated accurately using particle tracking if the subsurface is represented in such detail that mechanical dispersion (dispersion created by unrepresented heterogeneity) is explicitly represented and molecular diffusion is expected to be negligible. The effects of any unrepresented mechanical and molecular dispersion need to be considered when approximating arrival times using particle tracking or random walk methods. There have been some attempts to adjust advective transport to compensate for some of the problems discussed in this section; the piston exponential model of Cook and Böhlke (2000) is one example.

Water samples measured at one point in a groundwater system can have a complicated history, and care is needed when comparing simulated values like age to values derived from water samples. For example, well-bore flow (Clark and others, 2008) or other components of preferential flow or dual porosity (for example, Worthington, 2007) contribute to the pathways of flow. Consequently, calculated time-of-travel may be too long or too short because the model does not represent important aspects of the flow path. Things to consider include the geometry of flow (divergent versus convergent flow), flow mixtures from multiple inflows, preferential flow, or other alterations produced by field measurements of fractured or interstitial flow properties not represented at all or not represented accurately in the model. Time-of-travel observations can be equated to age dates for some types of field-derived geochemical samples. For example, simple monotonic decay or production of unstable isotopes with a relatively constant initial concentration such as carbon-14 or helium-4, respectively, combined with the assumptions of a homogeneous aquifers with steady flow conditions might permit groundwater transport to approximate a simple plug or piston flow conceptual model. Variable initial concentration history (fig. 1B), heterogeneous aquifers, and unsteady flow conditions complicate the analysis. MODPATH-OBS is designed to provide additional tools that enable modelers to more systematically make quantitative comparisons between particle-tracking results and age-tracer or concentration data that may be affected by mixing of waters of different ages and recharge sources.

MODPATH-OBS and Model Analysis

In addition to being used to generate simulated equivalents to transport-related observations, and for constraining predictions, MODPATH-OBS facilitates the use of more sophisticated model analysis methods. These model analysis methods generally address simple assessments of chemical and isotopic data relative to a simulated flow field, model fit to observations, sensitivity analysis (including data needs assessment), parameter estimation, and uncertainty quantification, as shown in figure 2.

To facilitate model analysis and output of observations from simulated flow fields, MODPATH-OBS lists simulated values in files defined by keywords XYZDAT, TIMDAT, CONDAT, and TYPDAT in the Output_Files input block of the MODPATH-OBS main input file, which is described in appendix A. In the same input block, keywords XYZINS, TIMINS, CONINS, and TYPINS can be used to create instruction files needed by commonly used model analysis programs PEST, PEST++, and UCODE-2005 to read the simulated values.

Using the simulated values from MODPATH-OBS, the programs PEST, PEST++, and UCODE-2005 can calculate sensitivities, which indicate how much each simulated value responds to a change in one parameter value. Sensitivities can be used to calculate the sensitivity and uncertainty measures listed in figure 2; these calculations are not computationally demanding. Methods based on sensitivities are called local or linear model analysis methods, and tend to require a number of model runs on the order of twice the number of defined parameters. Thus, for 10 model parameters, sensitivity analysis commonly requires about 20 model runs; in contrast, estimating parameters commonly requires up to a few hundreds of model runs. Local model analysis methods are discussed by Hill and Tiedeman (2007). Inverse methods using singular value decomposition (SVD) parameter transformation described by Aster and others (2012) and available in PEST also can require hundreds of model runs to estimate parameter values.

Model analysis methods that are advantageous when considering very nonlinear problems include global methods such as FAST (Saltelli and others, 2008), which generally require thousands to hundreds of thousands of model runs. Insight into the simulated dynamics dominated by different observations can be investigated using multi-objective function optimization described by Deb (2001) and available in Graphical User Interface for Multi-Objectives Optimization (GUIMOO) and PEST. Model fit can be investigated using objective function surfaces constructed for up to three parameters or combinations of parameters, as described by Hill and Tiedeman (2007, p. 78,82) and available in UCODE-2005.

Model Fit

- How to include many data types with variable quality?
  Error-based weighting, single objective function, MOO*

- Is model misfit or overfit a problem?
  Maximum-likelihood & bias-corrected variance

Sensitivity and Uncertainty

Observations ⟷ Parameters

- Which parameters can be estimated with the observations?
  b/SD_b, CSS&PDD, ID, FAST*

- Which observations are most important to parameters?
  Leverage, Cook's D, CV*, MOO*

- Are any parameters dominated by one observation and, thus, its error?
  Leverage, DFBETAS, CV*

- How certain are estimated parameter values?
  b/SD_b, Parameter uncertainty intervals[#]

Parameters ⟷ Predictions

- Which parameters are most important to the predictions?
  PPR

- How certain are the predictions?
  z/SD_z, Prediction uncertainty intervals[#], multi-model analysis*

Observations ⟷ Predictions

- Which observations are most important to the predictions?
  OPR, CV*

- Which of many conceptual models are likely to produce the best predictions?
  AIC, AICc, BIC, KIC (These statistics balance misfit and overfit)

Figure 2. Summary of model fitting methods and sensitivity and uncertainty methods that address the listed questions. Acronyms and symbols: MOO*, multi-objective optimization; b, parameter value; SD, standard deviation of the quantity in the numerator; b/SD forms a t statistic; CSS, composite scaled sensitivity; PCC, parameter correlation coefficient; ID, identifiability statistic for SVD parameters; FAST*, Fourier amplitude sensitivity testing; CV*, cross-validation; PPR, parameter-prediction statistic; OPR, Observation-Prediction statistic; and z, prediction. The sensitivity analysis and uncertainty evaluation methods are organized by what part they address of the observation—parameter—prediction triad. The model can be thought of as quantitatively connecting the parts of the triad if observations refer to simulated equivalents of the observations. *More computationally demanding statistics that are practical for models with very short execution times. #Uncertainty intervals can be calculated using frequentist or Bayesian approaches and linear, nonlinear*, Monte Carlo*, and bootstrap* methods. Figure modified from Hill and Tiedeman, 2007.

Calculation Methods

The four observations types are proximity, time of travel, concentration, and percentage of source water from contributing areas or other types of sources such as artificial recharge or contaminants. Concentrations can include age determinations based on nuclear decay (for example carbon-14 age) or production (for example helium-4) or age determinations based on a chemical decay (or production) rate. In MODPATH-OBS these simulated equivalents of observations from particle tracking are termed observations. In cases where these observations are linked to parameter estimation, sensitivity or uncertainty analysis, they are herein referred to as predictions.

Tracking directions, geometries, and measurement metrics applicable to the observation and prediction types are summarized in table 2. Table 2 also suggests how MODPATH-OBS capabilities can be used for observations and predictions. All observation/prediction types can be used with steady-state or transient flow systems, or flow systems that are sometimes steady and sometimes transient.

When execution times to complete a flow simulation are long and a free water-table surface is being simulated (called a convertible layer in MODFLOW), execution times often can be reduced by 50 percent or more by approximating the top of the system by using defined thicknesses layers (called confined layers in MODFLOW). This approximation was suggested by Hill (2006) and discussed in detail by Provost and others (written commun., 2011). The reduction in execution time is very helpful when many model runs are simulated for sensitivity analysis and regression. Observed heads can be used to approximate the elevation of the top of the saturated zone, and this can be used to determine a simulated top of the system. The elevation of the top can be updated as needed, and a free surface can be simulated as execution times allow.

Methods Applicable to All Observation Types and All Grids

The particle track for each observation is calculated independently of particle tracks for all other observations and predictions. Thus, there are no restrictions about combining different types of observations and predictions. In an embedded grid, all particles are tracked through the locally refined grid. In the area occupied by a local refined (child) grid, the particle tracking will be simulated by that child grid. The regional (parent) grid is replaced in those areas. In locally refined model grids, particles used in MODPATH-OBS can travel across any combination of parent and child models. Observations from child or parent grids can be used to estimate parameters from any grid.

For the Observation Process, MODPATH-LGR supplies global coordinates in all directions and includes an identifier for each particle to facilitate the grouping of particles by source or observation location. See appendix A for additional information on MODPATH-LGR for MODPATH-OBS.

Observations and predictions supported by particle tracking are characterized by the simulated beginning and end of one or more particle tracks. Particle tracking can be performed forward in time (so that the beginning position of the particle can generally be thought of as the source) or backward in time (so that the beginning position of the particle location generally can be thought of as the final location).

MODPATH-OBS allows the user to define locations for sources (forward tracking) and observations or locations for observations (backward tracking). These locations can be points, lines, areas, or volumes (see fig. 3A). The definition of areas and volumes is similar to the zone concept used originally by MODPATH (Pollack, 1989, 1994) and also employs the same definition of sources for particles within a model cell, as illustrated in figure 3B.

Points and lines can be defined anywhere in the model domain, although lines can only be vertical (for example, to define the screened interval of a well). They are defined by supplying the cell indices and a local coordinate (0 to 1) within the cell. Since MODFLOW is a cell-based model, the point locations should generally be used for locations that are not large sources or sinks of water. Examples of these might be an observation well (observation location) or a septic system that is not appreciably altering the flow system (source location). Given the cell based nature of MODFLOW, it is generally necessary to start points at these locations (backwards tracking for observation locations, forward tracking for source locations) for MODPATH-OBS to associate particles to those locations.

In general, for locations defined with shapes other than points, particles will be started or terminated at multiple points within the shapes. Depending on how particles are started, they may represent different volumes of water. For example, if the same number of particles were started on each face of a well cell for back-tracking, the particles from the varying faces would likely represent different volumes of water (flow across face/number of particles started on face) since the flows across the various faces are likely different. To account for this, volumes can be assigned to each particle using a column on the right-hand side of the endpoint file. This could be done by a program that runs between execution of MODPATH and MODPATH-OBS, or by using the label feature included with MODPATH-LGR and MODPATH6. The index of this column can be specified in the Options input block. If this option is not used, all particles will be considered to have the same volume and particles should be started accordingly.

Table 2. Observation types and basic characteristics.

[All source and observation locations can be defined as a point, area, or volume]

Observation type	Common particle tracking direction	Observation metrics[1]	Example of data needed to support an observation[2]	Example prediction
Proximity	Forward or backward	Proximity of particle(s) to defined location. Directional components or Euclidean distance.[3]	Concentrations of conservative solute movement over time interpreted to define proximity as a distance between observed and simulated center of mass or plume front at specific times.	Distance of plume movement from a source location[8] or proximity to a destination at a defined time.
Time-of-travel	Forward or backward	Time of travel between two locations along the flowpath of "transport."	Streamflow concentrations reflecting contribution to a stream reach from groundwater or solute from a defined source, interpreted to define first arrival[4] or arrival of the center of mass.	For a proposed landfill, how long is it likely to take for the center of mass of leakage to reach the outlet of the flow system?
Particle-concentration	Forward or backward	Concentration of particles.[5]	Concentrations of anthropogenic tracers at a well, interpreted to define water age.	Given a pollutant spill, how long will it take to reach a feature of interest?
Source-water contribution	Backward[6]	Percentage of all particles associated with an observation that are from a given source inflow location[7]	Concentrations of constituents derived from a user-specified source inflow location to describe mixing or contributions from multiple sources.	For a proposed outflow, identify potential contribution from sources such as recharge, underflow, or contributing areas of concern.

[1]If more than one particle is used, the simulated equivalent to the observations can be obtained as the minimum, arithmetic average, median, or maximum, of the noted metrics. Information for an exceedance curve can be produced. Exceedance is with respect to a user-specified threshold and is expressed as a fraction of exceedance between 0 and 1.

[2]The effects of dispersion are likely to be larger for more lumped representations of hydraulic conductivity and increase as actual heterogeneity increases. This needs to be accounted for in interpreting advective travel distance from concentration data. Please see description in Proximity section of report.

[3]Simulated values can be directional proximity components between the simulated and observed locations, including the x-component (parallel to the model row direction), the y-component (parallel to the model column direction), and the z-component (parallel to the model layer direction). Simulated values can also be the total difference in Euclidean distance (calculated as the square root of the sum of the squared directional distance components in two or three dimensions). Although one particle may be useful in some circumstances, more particles can be considered to evaluate the effects of small changes in initial particle placement and conditions along the path of travel. When more than one particle is used, the simulated distance(s) can be specified as a minimum, maximum, average, or median of the particle values.

[4]First arrivals generally can only be simulated accurately using particle tracking if the subsurface is represented in such detail that mechanical dispersion (dispersion created by unrepresented heterogeneity) is explicitly represented. The effects of any unrepresented mechanical and molecular dispersion need to be considered when approximating arrival times using particle tracking, and they generally are difficult to quantify.

[5]Concentrations are calculated as the number of particles from a particular source that occur at a user-specified location as a percentage of the total particles from that location. A large number of particles is needed to create reasonable approximations.

[6]MODPATH-OBS can be used with a large number of particles for forward tracking but may be more efficient for backward particle tracking for Source-Water Type Observations.

[7]Methodology modified from Sanford and others (2003, 2004).

[8]For this application the user needs to define the observation location to be the same as the source location.

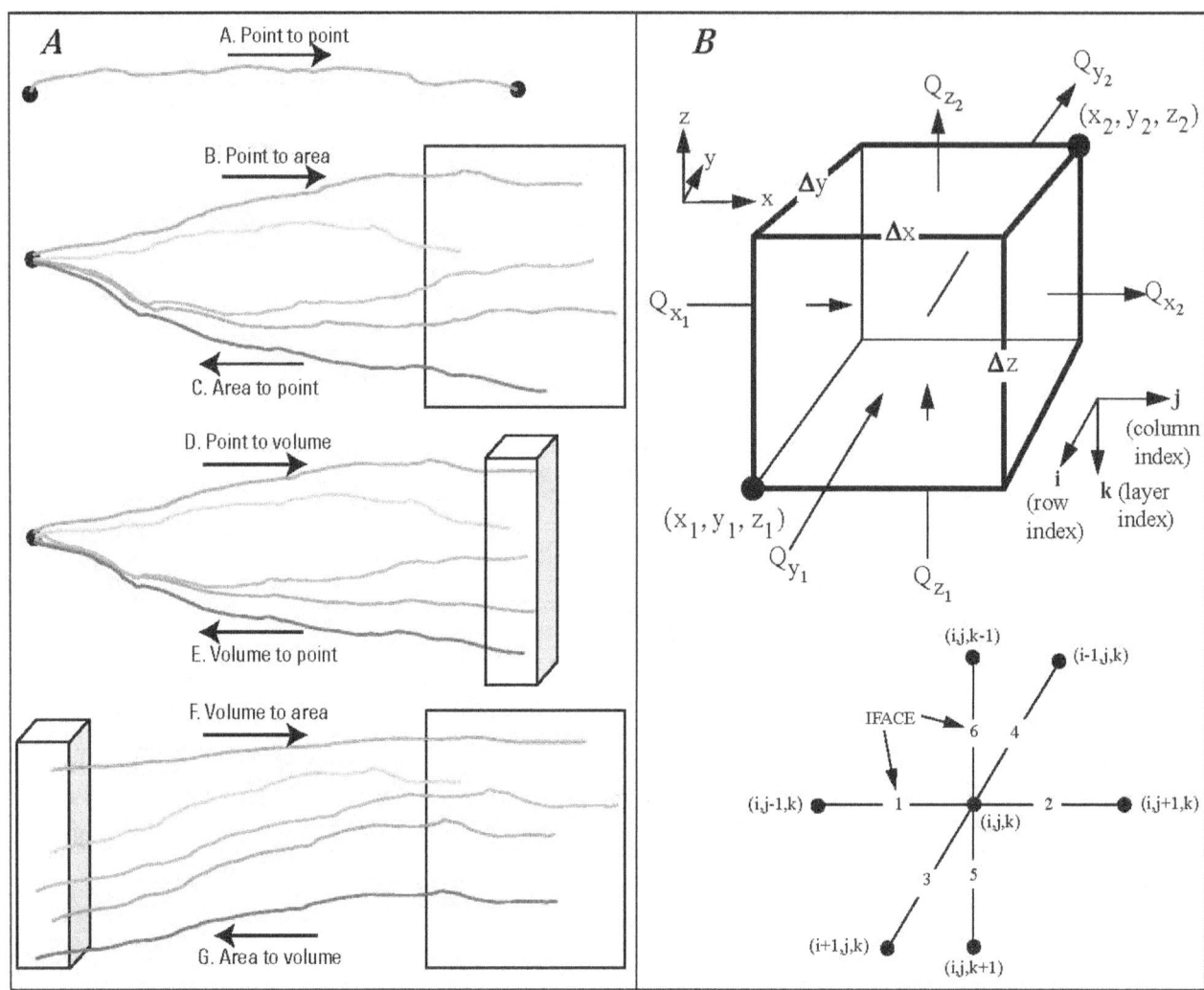

Figure 3. (*A*) Particle tracks may begin and end at points, areas, and volumes, and (*B*) local coordinate systems with respect to each model cell and numbers identifying the six cell faces of a cell, as used to define areas [figure *B* modified from Pollack, 1994. A value of zero designates all six cell faces.]

When volume information is included, it will be used for computing observation values based on multiple particles, except for those observations where a minimum or maximum is specified. If a median is specified, the value associated with the particle represents the 50th percentile of volume. When an average simulated value (Sim_value) is computed (as for concentration and when comp_ type equals the average for time-of-travel and proximity observations), the following equation is used:

$$\text{Sim\_value} = \sum(\text{particle value} * \text{particle volume})/ \sum(\text{particle volumes})$$

Source-water contribution observations would use the same equation where particle value equals 1 for the correct source location and particle value equals 0 for other source locations. Assigned volumes are assumed to be conserved along the flow path of each particle. So, if particles are passing through weak sinks or sources, the volume remains the same; the user should be aware of this assumption if this may have an effect on the simulated volume represented by the particles.

The value of the simulated equivalent generally is significantly affected by the arrangement of particles in the defined volume and by the averaging method. These are user-controlled and need to be carefully considered. MODPATH-LGR and MODPATH 6 allows users to define a volume using the LABEL column after the word "volume and evenly distribute particles within the volume or on the outside edge of the volume; this can be useful for some situations. The user-specified particle distributions will affect the simulated concentrations and the simulated fractions of particles used to estimate source types. Volumes and areas are defined based on model cells or cell faces. Volumes are simply defined by a subset of the model cells. They can be specified by supplying the cell indices or based on an array of values defining zones. Areas are defined in a similar manner, but with the additional specification of the cell face. Cell faces are defined using the conventions of MODPATH (figure 3B). These will generally be used for sources or sinks associated with fluxes at model boundaries. MODPATH allows fluxes to be associated with external boundaries when the auxiliary variable IFACE is specified in MODFLOW list-based stress packages (along with the COMPACT BUDGET AUX option in the output control file) and through the use of IRCHTP and IEVTTP for the array-based stress packages, recharge and evapotranspiration. The use of these options is encouraged to get meaningful results from MODPATH-OBS and to minimize the effects of weak sinks and sources in MODPATH. Similar to the point/line locations, if a source or sink is not associated with a volume/area location, particles should be started from those locations and backwards tracked for observation locations, or forward tracked for source locations.

Samples derived from water-supply wells pumped at operational pumping rates through depth-dependent sampling (Izbicki and others, 1999) have been successfully used to assess the complex flow and hydraulics near well fields in regional flow systems (Hanson and others, 2002). The ability to generate simulated observations and pair these with depth-dependent samples from supply wells, as well as the more common depth-specific samples from short-screened observation wells, provides a broader basis for comparison of regional and local-scale movement, concentrations, and mixtures of source waters and related dissolved constituents.

Methods for Locally Refined Grids

The location and coordinate systems used by MODPATH are also used within MODPATH-LGR and MODPATH-OBS (fig. 3). To define sources and destinations clearly sometimes requires designation of the cell face. In addition, a cell face

number of 0 (zero) is used to specify the entire cell, which is consistent with defining all cell faces.

Particles can be tracked within a single model grid or between a regional (parent) and one or more embedded, local (child) model grids (up to a maximum of 10 child models with MODFLOW-LGR, version 1.1). Particles tracked for advective-transport observations and predictions can begin in any of the defined grids and end in any of the defined grids. Particles are allowed to move freely throughout a system represented using child grids. Time and distance units used need to be consistent for all models used.

When calibrating a locally refined model, it is often useful to conduct preliminary calibration runs with advective transport observations in the parent model before proceeding to the child model(s). The resulting improvement of the parent model is likely to reduce problems in calibrating the more computationally demanding child model(s).

Proximity

The proximity observation type is used to consider transport from one known location to another known location. In other words, this type of observation is useful if the water quality indicates that at least some of the water at an observation location came from a known source location. In the case where the source is limited in time, the time of travel between the source and observations is also important and can be addressed by limiting the time of travel in MODPATH. When predictions are considered, the distance traveled over a specified time may be of interest, in which case proximity can be calculated relative to the starting location by also defining it as an observation location.

Description of Method

The method for proximity observations is best described using an example. Figure 4 shows an example from Anderman and others (1996) and Anderman and Hill (2001). At this site, the contaminant source was active long enough to produce an extensive plume. A number of monitoring wells indicate the extent of contaminated ground water. The sharp fronts of the concentration contours indicate that the movement of the contaminant plume is largely due to advection; the direction and length of the plume are controlled by the hydrogeologic characteristics of the field site. The configuration of the contaminant plume represents aquifer conditions integrated over the time of contaminant movement. A proximity observation can be obtained by contouring measured

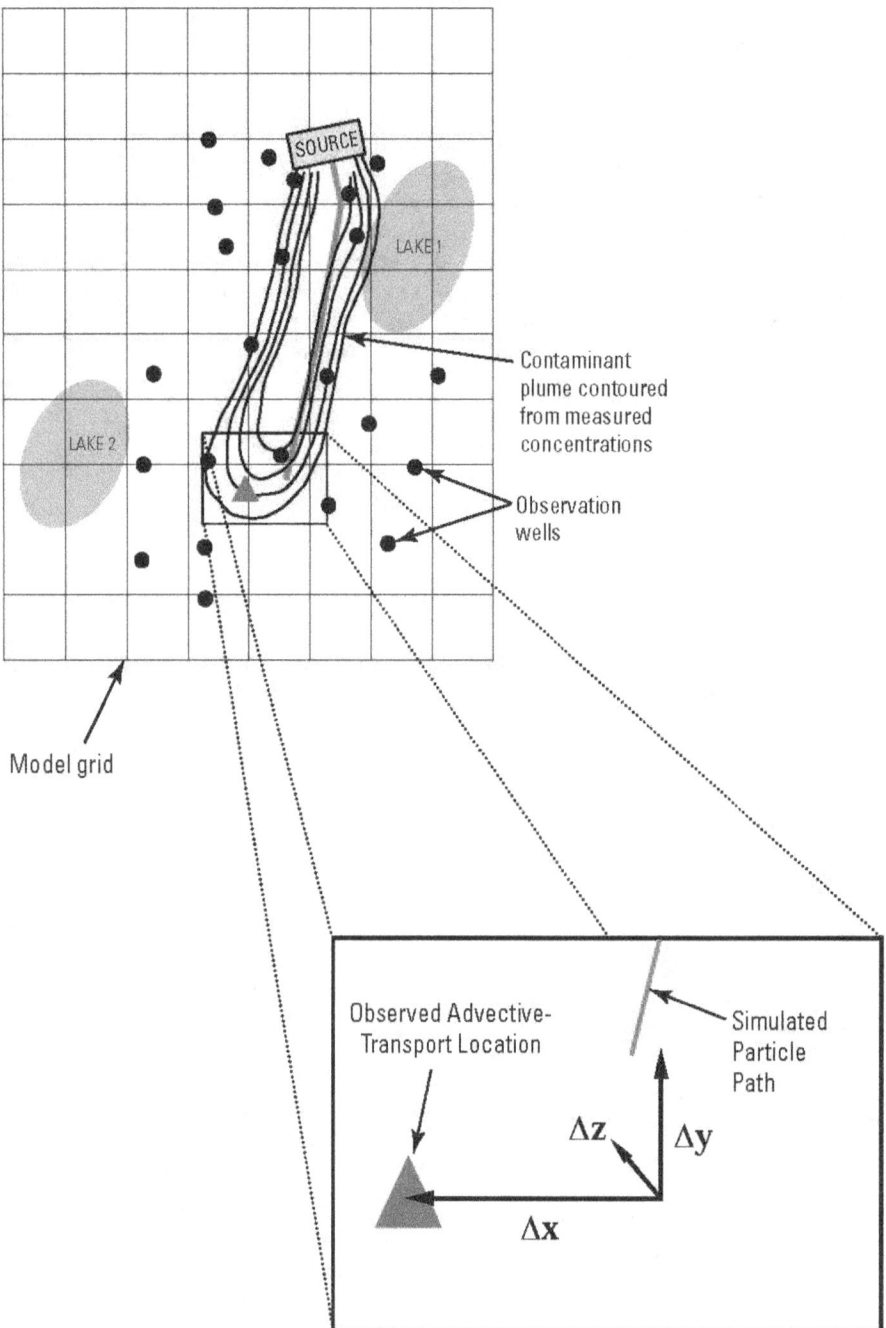

Figure 4.   Conceptual representation of distance and direction observations. [The cells are 1,500-meter squares. Contours show boron concentrations of 400, 300, 200 and 100 micrograms per liter. The source concentration is poorly defined and probably changed over time, and the approximate nature of the source contributes to uncertainty of the observation. The blue triangle shows the approximated location of the advective front of the plume, and can be used as the observation location for a proximity observation. This point then becomes the reference point for the observation, with a distance of zero. The red line shows a simulated particle path. The end of the path is compared to the observation location, and the total distance between them, or component distances in any of the three axis directions, can be used as simulated values. Figure modified from Anderman and Hill, 2001.

contaminant concentrations and choosing a suitable point along the plume front, as shown by the triangle in figure 4. The selection of the advective-front location is generally not straight forward as was discussed earlier in "Relating Simulated Advective Transport to Field Conditions". Once an observation is obtained, a simulated equivalent is needed.

To simulate advective transport, a particle can be forward tracked through the grid from the source location for a specified length of time, as shown by the red line in figure 4. Backward particle tracking can also be useful, for example, when identifying source locations; a particle would be introduced at the final position of the plume front (called the observation location) and tracked backward toward the source. In this situation, the reference time would be the time when the particle was thought to have arrived at the observation location. For example, in the hypothetical example problem, an arrival time of 2010 is used and the backward particle tracking counts backward in time from 2010.

MODPATH-OBS is programmed such that for proximity observations, the observed value is zero. The simulated value of proximity equals the distance of a particle's final location to the observation location for forward tracking (particles started at the source location) or to the source location for backward tracking (particles started at the observation location). The proximity can be reported as the Euclidean distance from a particle to the desired location or a directional component (x, y, and z) of the distance. The proximity distance (D) is calculated as $D = \sqrt{x^2 + y^2 + z^2}$. To use proximity distance for predictions of the distance from the source, the observation location remains the same but backward versus forward tracking can be used and the "observed" location is set equal to the source location. To accomplish this, an observation location that is the same as the source location needs to be added in the input file.

As indicated in table 2, proximity calculations can be conducted for one or many particles. It might be useful, for example, to include the effects of slightly different source locations on simulated proximity. Results from the different points can, be averaged to produce the simulated value of proximity. Alternatively, the median, maximum, or minimum may be chosen. The minimum generally should be used, since that is what is best supported by observed water-quality data. The observation supporting this type of observation would likely be that some constituent originating from a known source area is found at an observation location. All that can really be said is that at least some water got from the source to the observation location.

Proximity observations are the only observation type defined in this report for which a non-zero sensitivity is calculated even when the flow field does not transport any particles to the intended destination. Thus, proximity observations may be useful at the beginning of model development when the simulated flow field may be quite

different than is consistent with what is known about transport. Once the simulated flow field is oriented in the right direction through the model calibration process, time-of-travel observations become increasingly important because the proximity observations may have zero sensitivity. However, because the different parameter values used in parameter-estimation methods may again result in very different simulated flow fields, it is worthwhile to retain the proximity observations throughout the model calibration process.

A problem that can occur with proximity observations is that the particles may exit the system via wells, streams, or some other stress or boundary condition before the time of the observation. This is a problem because if a particle has gone farther than expected, we would like the regression to be so informed, enabling the appropriate adjustments in parameter values to get the particle to the right location at the right time. To achieve this goal, particles that leave the system can have attributes of time or location projected

Proximity Observations when Using Areas and Volumes

When areas or volumes are used to define a particle destination, proximity is defined as the distance to the closest point on the volume or surface of the location model cell, as shown in figure 5. Anywhere within the area or volume is assigned a distance coordinate of (0,0). Also, each axis has zero values over a segment of the line in each axis direction rather than at a point as would be the case in a standard coordinate system. In these cases, proximity is a composite relative difference in distance. However, for estimation of individual components (x, y, or z), the location components are still relative to the zero location of the observed value.

Proximity Observations and Weighting

The importance of proximity observations (and other types of observations) that result in simulated predictions or comparisons with field data during parameter estimation, uncertainty or sensitivity analysis will partly depend on the user's application of a weighting scheme for the observations used in the analysis. We again turn to the example in figure 4 to illustrate the weighting of proximity observations. The source is a sewage-discharge plume at Otis Air Force Base, Cape Cod, Massachusetts (LeBlanc, 1984a,b). The analytical solutions shown in figure 1 were used to determine measures of advective travel from concentration observations subject to dispersion, and to evaluate the likely accuracy of those measures. In the analytical solutions, the source size was varied from 600 to 1,200 ft and the transverse dispersivity was varied from 13 to 30 ft, as suggested for this site by LeBlanc

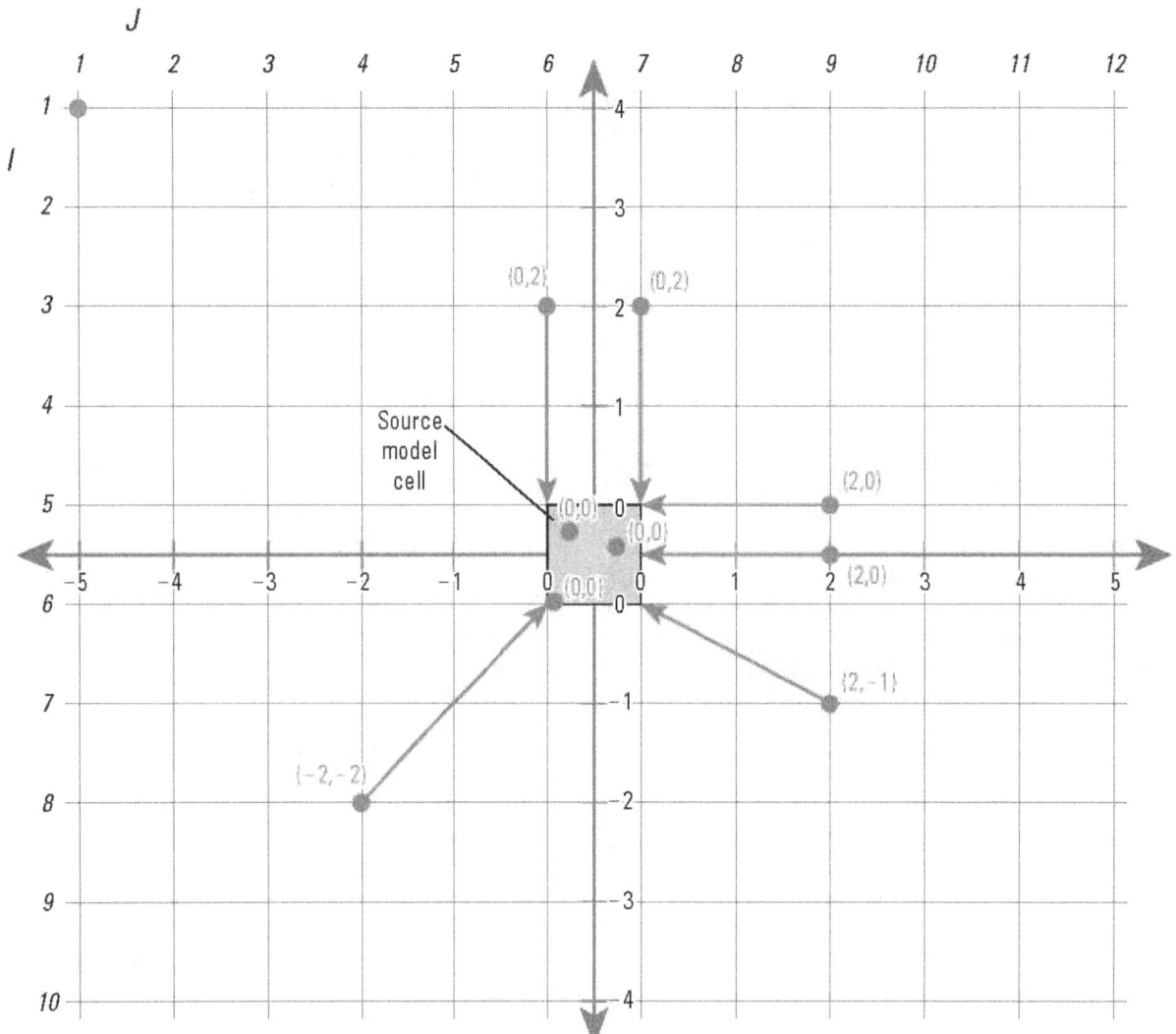

Figure 5.   Directional distances (shown in red) from a rectangular area defined as a single model cell. [Within the blue box of the model cell of the location the proximity distance equals zero. Use of a source in this figure suggests one of two options. First, the particles could be tracked backwards from an observation location and the proximity of the particle to the source is of interest. Second, the particles are being forward tracked and distance traveled from the source is to be calculated. This requires defining the observation location coincident with the source location. Thus, for forward tracking, the distance reported for proximity observations is always relative to the observation location shown by blue arrows (shortest distance for each particle between observation and cell faces of source cell); for backward tracking the distance is always relative to the source location.

(1984a,b). Results suggest that the normalized concentration at the plume advective front is between 17 and 43 percent.

Figure 4 shows that the contours are not well defined by concentration measurements. Also, as is common, the source concentrations are not well defined. In this situation, the front of the plume is not thought to be very well defined. It was thought that the location could easily be in error by as much as 500 meters in both the X and Y direction of this two-dimensional model. This statement can be quantified if a probability distribution and significance level are assumed. If a normal probability distribution is assumed, and plus and minus 500 are assumed to have a 65 percent probability of including the true value, a standard deviation of 500 m results.

Table 3 lists the coordinates of the starting location of a single particle used in this problem, the associated observation location and distance traveled in each direction given the final calibrated parameter values, and the standard deviation of the distance used to calculate the observation weighting. The weight is calculated as the reciprocal of the variance; the variance equals the square of the standard deviation. Weighting determined by an analysis of error is called error-based weighting by Foglia and others (2009), and is consistent with the ideas discussed by Hill and Tiedeman (2007, p. 291–305). Error-based weighting is advantageous because theory shows it produces the parameter estimates with the smallest variance. Also, many commonly used measures of uncertainty depend on the weighting being error-based. The proofs for both of these characteristics of error-based weighting are shown by Hill and Tiedeman (2007).

For proximity observations, simulated equivalents are obtained by transporting one or more particles for specified lengths of time. Particle movement is recorded at each time in the pathline or time-series file, and the total distance traveled by each particle is recorded in the endpoint file. Movement in any combination of the three coordinate directions—x (along rows), y (along columns), and z (vertical), or total rectilinear distance can be recorded. If particles stagnate or do not leave the initial model cell they are not counted as checked against the termination code indicator (IPCODE) from MODPATH. If the particle leaves the model before the defined time or end of total simulation time, one of three things can occur, depending on chosen input file options. First, the particle can be propagated at the velocity projected from its point of exit from either a parent or child model if the IPCODE is greater than zero and the ADVOBS option has been specified with MODPATH-LGR. These projected particle endpoints need to be appended onto the global endpoint file. Second, the particle can have a normal exit or termination point within one of the models (IPCODE =1,2). Thirdly, the particle can never be released or is stranded in an inactive (dry) cell (IPCODE=-1,-2). Distance observations will be estimated for particles that discharge normally or stop in a specified zone of observation (IPCODE = 1, 2).

Table 3. The coordinates of the starting location of the single particle used in the problem shown in schematic form by figure 4, the associated observation location and distance from the starting location to the observation location in each direction, and the standard deviation of the distance used to calculate weighting in the regression.

[Values are in meters. Modified from Anderman and Hill (2001, p. 28)]

Grid direction	Starting location	Observation location	Distance (Obs-Start)	Standard deviation
X	7,750	7,000	-750	500
Y	8,625	16,500	7,875	500

Time of Travel

The time-of-travel observation is used to consider the time of transport from one defined location to another. The same approach is used when observations or predictions are considered. Information pertaining to the time and path of travel can be determined from a tracer test or from apparent age determinations of water samples that originate from a known location. In addition, groundwater velocities can be estimated directly using data from in-situ temperature probes (Ballard, 1996), indirectly using temperature perturbation methods (Constantz and others, 2003), or by particle-tracking three-dimensional well-bore flow, constrained by well-bore measurements (Newhouse and Hanson, 2000, 2002).

Description of Method

For time-of-travel observations, simulated equivalents are obtained by transporting one or more particles for specified distances or until they reach specified internal boundaries or volumes, or model boundaries. The time-of-travel is computed by subtracting the particle release time from the time when each particle reaches its final location. The time-of-travel is converted from the MODPATH time units and reported in the units specified in the options input block by keyword Time_Units.

One or more particles can be used for each time-of-travel observation. For example, multiple particles can be used to investigate the effects of different starting locations on the time-of-travel, especially for sources or sinks that create divergence or convergence of flow. If many particles are used, the simulated equivalent to the observations can be defined to be minimum, maximum, average, or median of the associated particles. In addition, the probability of exceeding a user-specified time can be determined, or exceedance curves of simulated values can be produced. Any location that represents two- or three-dimensional objects should be represented by more than a single particle.

Exceedance Curves

Data for producing exceedance curves can be produced by MODPATH-OBS (see table 1). Example exceedance curves (fig. 6) for particle tracking in models of Avra Valley, Arizona. (Zimmerman and others, 1991) show a probability of 1.0 for groundwater travel times longer than about 1,700 years, and a probability of 0.0 for that groundwater travel times longer than about 3,200 years. This model is an example of particle-track observations generated using Latin-Hypercube Monte Carlo perturbations of a transmissivity field.

Exceedance curves can be used to present results from time-of-travel calculations conducted using many particles. Zimmerman and others (1991) and Hanson and others (1996) used exceedance curves to present results from a number of model realizations (fig. 6). This example uses exceedance curves to show results of different transmissivity fields and boundary conditions from a study of Avra Valley, Arizona that represents simple flow paths within a regional-scale alluvial aquifer flow system (Zimmerman and others, 1991). These exceedance curves are obtained directly from particle-tracking: the curves count the particles as they arrive at the defined destination so that the early time records the arrival of the first particle and the last time records the arrival of the last particle. They show the larger difference in span and median time-of-travel when the realizations of transmissivity used in the Monte Carlo simulations are not constrained to prevent simulated heads above the land surface, which was not a plausible outcome. The curves obtained from this example also are influenced by the conceptualization of the flow system and censorship of flow-field realizations.

Generally, many particles are required to produce an exceedance curve for complex flow geometries and boundary conditions or regional-scale problems. Exceedance probability calculations are most applicable when the simulated time series is complex because of the arrival of many particles over time. Exceedance curves are most often produced with results from time-of-travel and concentration observation types, but their use is not restricted to these types of observations.

Missing the Location

There are cases where there might not be particles associated with an observation. Depending on the nature of the time observation, the time of travel can be limited to a single source location. In this case, even if particles were backtracked from the observation location, there is the possibility that no particles are associated with the desired source. In looking at point-to-point cases, the time-of-travel type observation is unlikely to be useful. For point-to-points cases where time of travel is important, the particle can be stopped at that time in MODPATH and a proximity observation used.

In the case where no particles are associated with a time-of-travel observation, no time can be calculated and MODPATH-OBS will print a user defined number in the output (NoPartValue) instead of a simulated value. The number can be used as follows.

1. The OmitDefault option of the UCODE-2005 Reg_GN_Controls input block can be used to omit this observation from regression calculations; similar capabilities are available with other model analysis programs. While this avoids performing calculations with the assigned default value, it does not solve the problem that the simulated velocity field is not producing the expected transport.

2. Replace the result with the result of another particle using the program Sim_Adjust (Poetter and Hill, 2008).

3. Convert the time-of-travel observation to a proximity observation, either permanently or until the flow field is simulated in a way that will transport the particle such that it is useful to the time-of-travel observation.

Concentration

Concentration refers to the concentration of a constituent for a collection of particles estimated at the observation location. This observation type allows for assignment of a source concentration for a particle, which can then degrade or accumulate during particle tracking. The volume-weighted average concentration of all particles contributing to an observation is used to define the concentration at the observation. If volume is not assigned on a per particle basis, the assumption is made that each particle represents the same volume of water. Given this, if not assigning volumes to the particles, the user needs to design the initial particle distribution such that each one represents a similar amount of water.

Concentration observations can be used to simulate the advective transport of anthropogenic tracers ranging from contaminants to unstable isotopes and other conservative tracers with known decay (or accumulation) properties (zero- and first-order equations are supported). Many processes that are important to transport (for example, dispersion) are not represented; if these processes are important to include, a more rigorous transport code is required.

For a time series of data, the concentration exceedances represent the spatial and temporal distribution of the observations that exceed the user-specified threshold value. For steady-state, this is computed from the particles used and calculated at equal intervals of time. For transient-state, particles will need to be released in conjunction with each observation. Therefore, if backward tracking, the user will release a pulse of particles at each time the observation location is sampled.

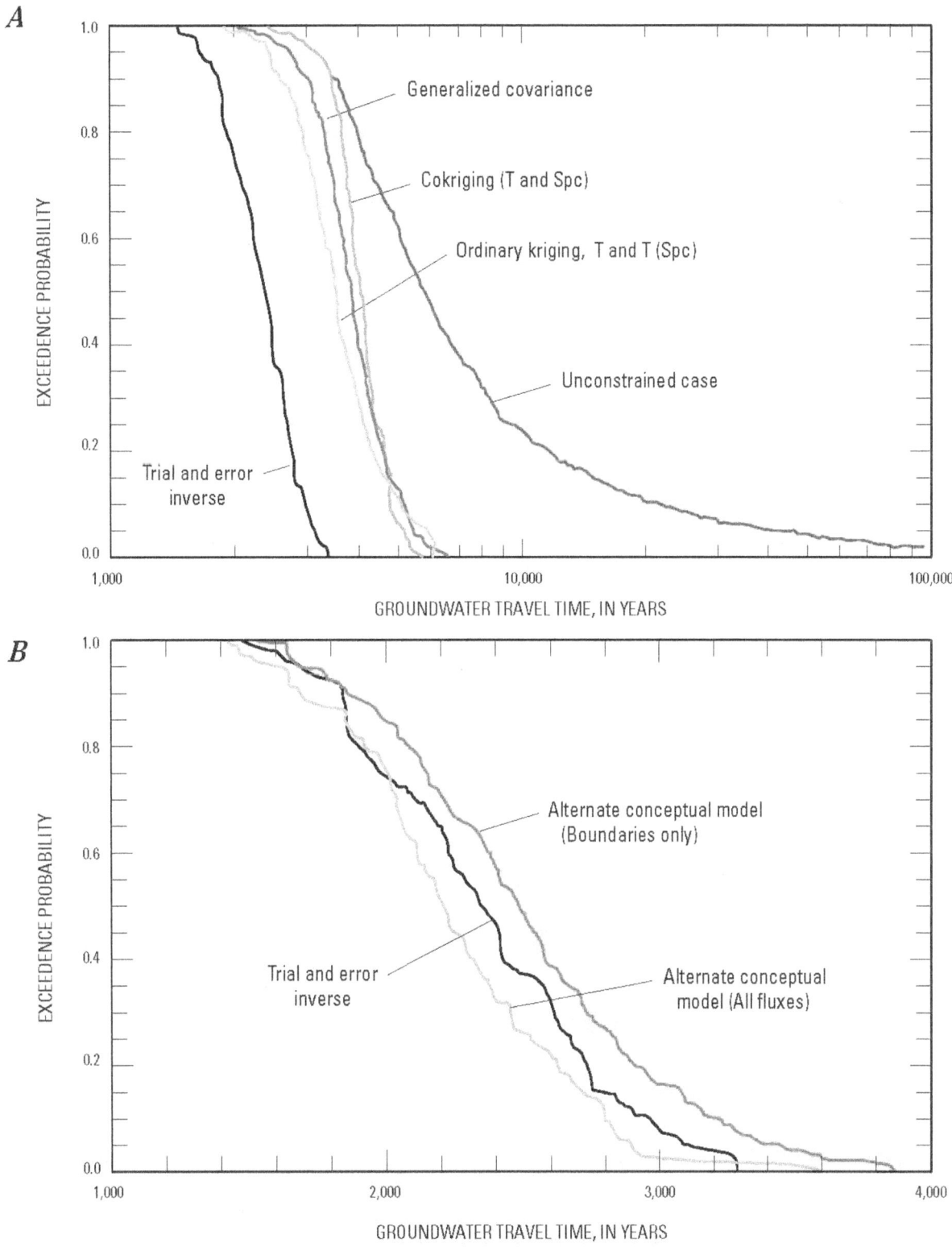

Figure 6.   Exceedance curves for simulation scenarios of particle-tracking travel times (modified from Zimmerman and others, 1991); (*A*) Results are shown for models with different transmissivity (T) distributions. "Unconstrained case" means that unrealistic cases in which the water table is simulated above land surface are included; (*B*) Different conceptual models with same transmissivity distribution. [MODPATH-OBS can produce output defining the value for one exceedance probability or an entire curve. The latter is attained by defining many observations with different exceedance probabilities. This figure is discussed under the time-of-travel observation type. T(Spc) are transmissivities derived from specific capacity tests.]

Description of Method

In developing particle-concentration observations with MODPATH-OBS, simulated equivalents are obtained by transporting one or more particles between an observation location and a source location. The travel time of these particles is matched with a historical profile of concentrations from user-specified time-series of initial concentrations at a time that is equal to or greater than the sampling time. Thus, the travel time is the elapsed time between the initial simulation time and the sampling time. The simulated observations are thus derived from the user-specified source concentration history and travel times of one or more particles; these concentrations are averaged with equal weighting, or volume-weighted, if specified by the user.

The MODPATH endpoint file is used to obtain the source location and travel time for each particle associated with a concentration observation. The time of recharge is calculated by subtracting the time-of-travel from the time at which the observation was made (SampleTime). Then the recharge time for the particle is compared to a user-defined history of concentration at the source where the particle was started – often the concentration of an environmental tracer, such as that shown in figure 1B. The initial concentration is determined by linear interpolation of the concentration history, or is set to the first or last value if the recharge time is outside of the user-supplied range. If specified by the user, the initial concentration can be degraded based on the total time of travel and a user-specified degradation (or accumulation) coefficient and equation. The resulting concentration is assigned to the particle at its destination; if no degradation (or accumulation) is specified, the initial concentration is assigned to the particle at its destination. Generally a large number of particles are used and the concentration at the destination is obtained by averaging the concentrations of the individual particles.

The order of decay and related decay rate, or a constant production rate, can also be specified for each type of concentration observation along with its historical time series. Nitrate degradation, for example can be simulated as a zero-order decay (Green and others, 2008) or first-order decay (McMahon and others, 2008) process. A constant production rate could be used for helium dating methods for old waters, defined using a negative decay rate (production rate).

MODPATH-OBS users can specify zero-order or first-order decay, which are calculated by MODPATH-OBS using the following equations.

$$\text{zero-order decay: } C = C_0 - kt \qquad (1)$$

$$\text{first-order decay: } C = C_0 e^{-kt} \qquad (2)$$

where

C_0 is the concentration at the initial time, as determined based on the time of recharge and the concentration history,

k is the decay (positive value) or production (negative value) constant, and

t is elapsed time since the particle was at the source location.

For backward tracking, the time decreases as the particle is tracked back to the source location. The decrease of concentration from decay stops if it reaches zero concentration. MODPATH-OBS also is able to track breakdown products that are controlled by a user-specified ratio between the parent concentration and the concentration of the breakdown product. One example is a stoichiometric tritium to ^{3}He ratio determined by the decay reaction.

Obtaining Particle-Concentration Observations

While presented in the context of concentrations, some types of age observations can be derived from these concentrations for monotonically decaying unstable isotopes (for example ^{14}C) or accumulating isotopes (for example ^{4}He), and as such, have a wide range of potential applications. In these cases, the equivalent concentrations can be transformed back into ages, in years before the sample collection time, by the user outside of the MODPATH-OBS program. Many other types of age date estimates derived from concentrations of constituents have predictable historical global or local concentration histories. Some of these constituents or isotopes have a monotonic history of decay or accumulation. For example, tritium/^{3}He, is a ratio that changes monotonically with age, sulfur hexafluoride (SF$_6$) has consistently been increasing in the atmosphere, and ^{4}He accumulates over time. Some other constituents, such as tritium and CFC (chlorofluorocarbon), have historical peak concentrations that may result in non-unique equivalent concentrations over periods of historical observations that span these peaks. In addition, the observed data may represent a mixture of waters with different ages, so it may not be advisable to convert these concentrations back to actual ages even when the concentrations are monotonic with age. As shown by Sanford and others (2003, 2004), concentrations such as percent modern carbon (as opposed to converting to age dates) are recommended for application of these concentrations during parameter estimation because they will result in a smoother distribution of estimated values and no discontinuities.

Similarly, for CFC and SF_6, it is recommended to convert concentrations measured in groundwater back to the "air equivalent concentrations" to eliminate issues associated with corrections for excess air, recharge temperature, and barometric pressure (see Crandall and others 2009) since MODPATH-OBS cannot account for these factors. In this case, the concentration provided in the concentration file will be the concentration in air.

When considering constituents that have concentration histories that are not smooth, it may not be feasible to adjust model parameters to the simulated concentrations across local observed maximum concentrations that occur within the time period of these concentrations to a value closer to the observed concentration. An example of this might be tritium which has multiple large spikes as well as seasonal variability. Smoothing the input curve may help the parameter estimation in such a case and the more accurate curve can always be put back in when the model is calibrated or close to it. Care should be taken in the application of these types of observations. They should be used in conjunction with other age tracers and other types of observations when possible to constrain their potentially non-unique outcomes in the comparison process. Usually even having two age tracers is enough to eliminate or reduce the non-uniqueness problem.

Converting ages to concentrations, such as ^{14}C activities, tritium units, and CFC concentrations, can help provide a smooth distribution for parameter estimation of estimated ages from observations derived from particle tracking. Sanford and others (2003, 2004) provide a good example of using ^{14}C age dates and transforming them into concentrations for smoother distributions during parameter estimation (figs. 7 and 8).

Cook and Böhlke (2000) divided subsurface transport models for environmental tracers into two general categories (IAEA, 2006):

1. Models that predict variations in ages within the aquifer; since ages increase with depth in most aquifers, these are referred to as "groundwater stratigraphy" models;

2. Models that predict the integrated age of water discharging from the system.

The first category refers to estimation of concentrations and related ages at specified locations within the flow system, and the second refers to integrated or mixed samples along regional flow paths that are observed at discharge points such as rivers, springs, or artesian flowing wells. The concentration curve data for environmental tracers used to date groundwater from years to decades using CFC, SF_6, ^{85}Kr, helium, and tritium-helium can be obtained from the Excel worksheet, TRACERMODEL1 (IAEA, 2006) or from

Tracer LPM (Jurgens and others, 2012). These are required to build the age tracer history for simulated observations that are matched to the location and time of the samples being used for observations.

Sanford and others (2004) provide an example of the second category, where the integrated ages and changes with depth provide a form of groundwater stratigraphy (figs. 7 and 8). Eberts and others (2011) employed techniques similar to MODPATH-OBS, using convolution integrals to provide particle-based estimations of average concentration (age-date) distributions to assess the vulnerability of production wells to contamination.

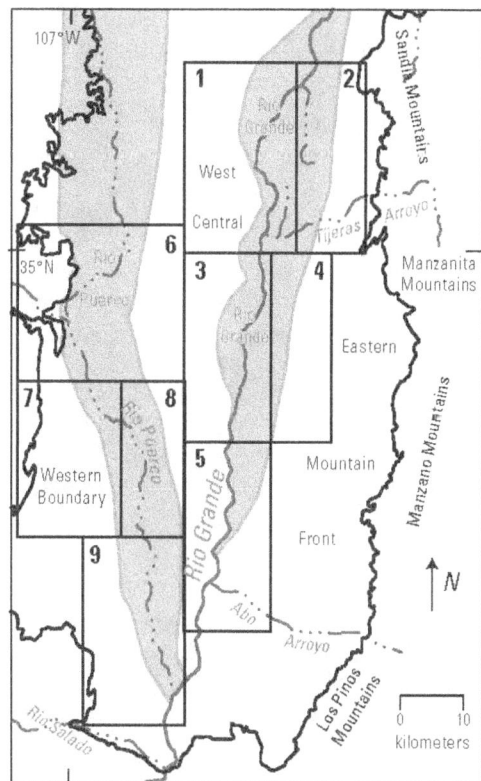

Figure 7.   Middle Rio Grande Basin groundwater flow model and source location regions 1 to 9 used for hydrochemical source observations. (Sanford and others, 2004)

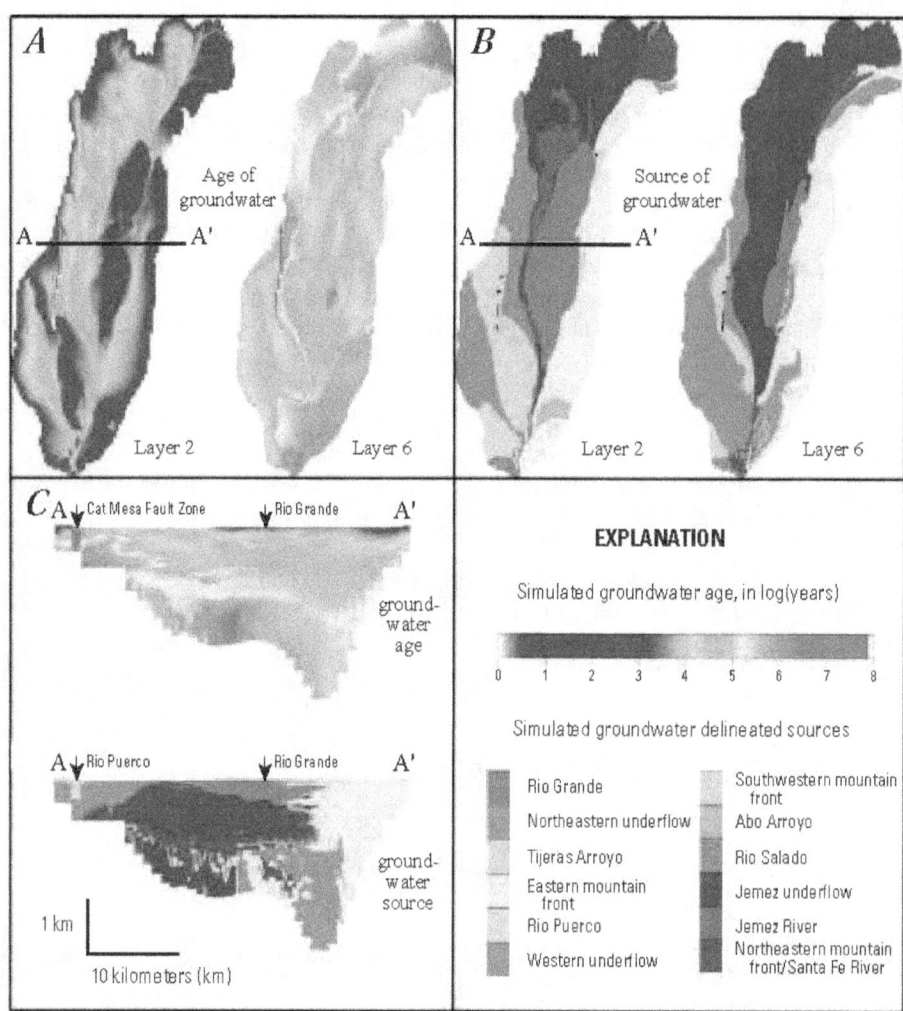

Figure 8.   Middle Rio Grande Basin model simulated results, (*A*) Groundwater ages in model layers 2 and 6; (*B*) Groundwater source-area delineation in model layers 2 and 6; (*C*) Groundwater ages and source area delineation along east–west section A–A'. (Sanford and others, 2004)

Source-Water Type

The source-water observation type is designed to simulate contributing percentages of source-water types to points, lines, areas, or volumes of interest. Source waters can be delineated for distinct areas or there can be multiple types from the same region, such as sources of natural and artificial recharge, or coastal inflow of seawater and injected water from barrier wells.

Description of Method

For source-water observations, simulated equivalents are obtained by backtracking one or more particles from an observation location to one or more source locations. The observations are expressed as fractions of particles from the specified source at any observation location. The number of particles used will directly relate to the precision of the source fractions; the user should consider the desired precision when determining the starting number of particles. For example, if 100 particles are started, the fraction can only vary in increments of 0.01. In the case where sensitivities are being calculated by perturbation (for example by UCODE or PEST), this can be important, especially if the sensitivities are ultimately used for parameter estimation.

One way to estimate percentages of source water from contributing areas is to use particle tracking at both regional (figs. 7, 8) and local scales (figs. 9, 10). MODPATH-OBS supports this type of comparison through calculation of the percentage of particles from user-specified source areas that reach a given observation location, and associated travel time(s). Source-water observations can have one or multiple values over time. For example, they can be specified for single or multiple times to accommodate the possibility of changes in natural or anthropogenic sources of water. The current version of the program does not allow for changing source over time; any changes in the source-water observations over time will be from changes in flow based on transient models. MODPATH-OBS uses particle IDs to define sources and calculate percentages of sources to support mixing estimates on the basis of geochemical data, such as ratios of anions (chloride, iodide, boron, and bromide), ratios of cations (calcium, magnesium, potassium, and sodium), or deuterium-oxygen isotopes.

Obtaining Source-Water Observations

Percentages of source water is another type of observation that can be used based on selected geochemical indicators to quantify the contribution of natural and anthropogenic sources to flow at points of observation within a groundwater flow system. As stated by Franke and others (1998), "Most human-derived contaminants in ground water are related to activities at the land surface and enter the ground-water flow system at the water table after passing through the unsaturated zone. A second important location of contaminant entry, which is of much smaller areal extent than the water table, is the beds and banks of streams, reservoirs, lakes, and wetlands. Given that most human-derived contaminants enter the groundwater flow system directly or indirectly from the land surface, one approach for protecting public groundwater supplies is to estimate areas contributing recharge to public-supply wells and then to implement groundwater protection practices on the overlying land surface." For example, source water from areas contributing recharge (previously called capture zones) for pumping wells and boundary conditions representing local or regional outflows such as springs and rivers can be delineated using particle-tracking simulations. Areas contributing recharge to wells, derived from transient and steady-state flow fields, often are used to assess vulnerabilities associated with given amounts of pumping.

Estimates of percentages of source waters can be used with observations of stable isotopes of deuterium or oxygen to estimate binary mixtures. For example, this approach was used to estimate the extent of artificial recharge in Santa Clara Valley, Calif. (Muir and Coplen, 1981, Newhouse and others, 2004), and the approach has been used to estimate higher-order mixtures from multiple sources (Plummer and others, 2004; Sanford and others, 2003, 2004). However, the estimation of source-water mixtures from stable isotopes needs to be carefully implemented by constraining the isotopic data by other geochemical data, such as carbon-14 ages. For example, the stable isotopic values from deeper wells in the Santa Clara Valley, Calif. actually represent lighter values associated with the last glacial period (> 17,000 years before present), and not recent snowmelt from the Sierra Nevada, the source of recently applied artificial recharge (Hanson and others, 2002). Thus, source-type observations must be carefully constructed to correctly represent the relation between source types and flow paths, which are controlled by the boundary conditions and aquifer properties that are being estimated during calibration or parameter estimation.

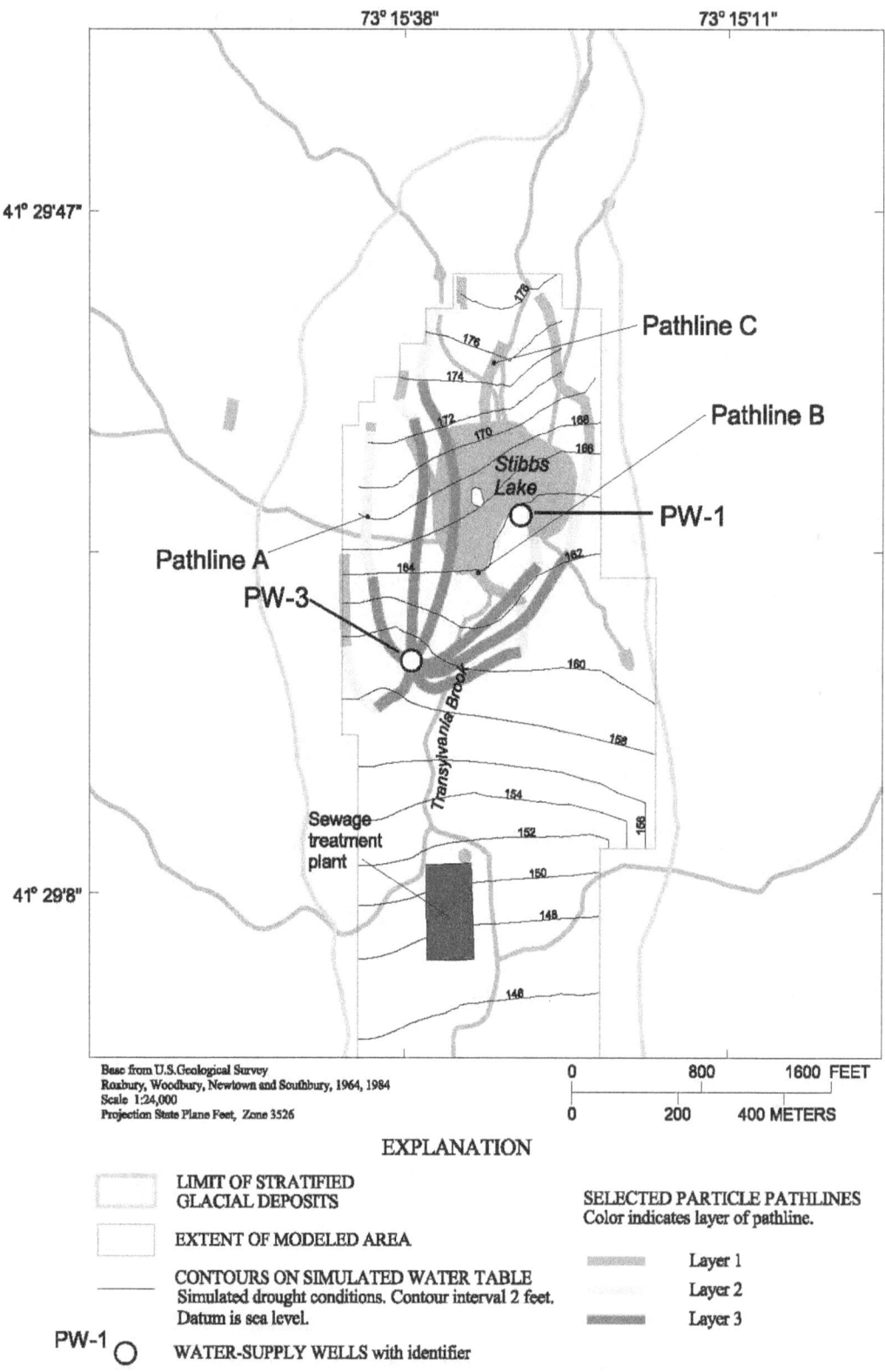

Figure 9.   Selected pathlines to a well at Southbury Training School, Southbury, Connecticut region. (Starn and others, 2000)

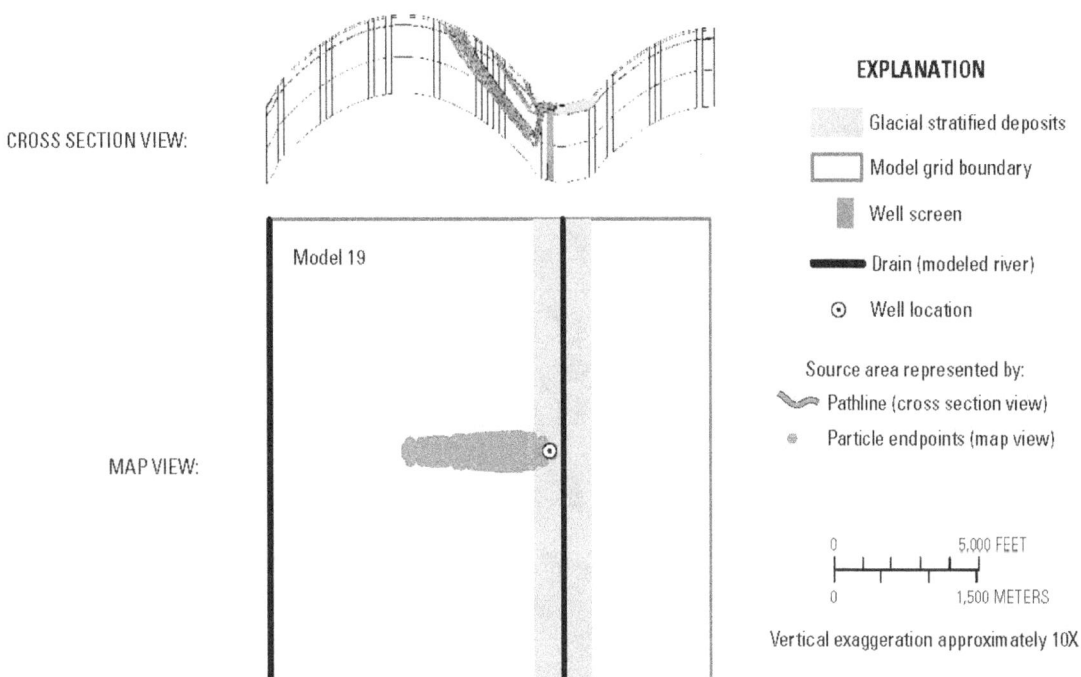

CROSS SECTION VIEW:

MAP VIEW:

Model 19

EXPLANATION

Glacial stratified deposits

Model grid boundary

Well screen

Drain (modeled river)

⊙ Well location

Source area represented by:

Pathline (cross section view)

Particle endpoints (map view)

0   5,000 FEET

0   1,500 METERS

Vertical exaggeration approximately 10X

Figure 10. Source-water areas to a well completed in bedrock in a narrow valley with a surficial aquifer, Connecticut. (Starn and Stone, 2004.)

In a broader context, the use of particle tracking, geochemical tracers, and flow models can help to identify contributing source waters as demonstrated in figures 9 and 10. In this context the notion of "contributing areas" is a special case of a broader class of source regions that include natural and anthropogenic indicators of source-water flow at a user-specified observation location. Thus, the class of source-water types include not only flows from the land surface (previously referred to as "contributing areas" or "capture zones") but also can include flows from other sources, such as groundwater underflow from adjacent basins, streamflow infiltration, subsurface adjacent aquifers and

confining units, faults as flow barriers or conduits, aquifer-storage-and-recovery systems, and multiple-aquifer well-bore flow. An example of an approach to estimating percentages of capture by superposition flow simulation was demonstrated by Leake and others (2008). Instead of using groups of particles, they used a budget analysis of simulated effects of groundwater withdrawals and artificial recharge to estimate the effects on discharge to streams, springs, and riparian vegetation in the Sierra Vista subwatershed of the Upper San Pedro Basin, southeastern Arizona. However, this is approach does not give information on water sources that can be obtained from the particle-tracking approach.

Common Calibration Problems

A number of problems are commonly encountered when using advective-transport observations in regression. Some of the most common problems are discussed here. The use of advective-transport observations may not be appropriate for all situations. Similar problems can occur for predictions.

1. Problem: The particle track at the starting parameter values is different than the observed track.

    Discussion: Commonly a particle will track in a different direction than the observed path from field observations because the simulated path is dependent on the starting parameter values. Come starting parameter values may even cause the simulated particles to exit the model grid prematurely. This difference can produce smaller than expected sensitivities that do not reflect the true value of the data.

    Resolution: Often this problem resolves itself. As the parameter values are changed by the regression during the parameter-estimation iterations so that they are more reflective of the actual system, the simulated particle track generally will also start looking more like the observed track. If it does not resolve itself and the final track is incorrect, see Problem 2.

2. Problem: Simulated equivalents at converged parameter values are very different from the related observations.

    Discussion: Possible causes include the following.

    (a) Particle entry point(s) are incorrect,

    (b) One or more of the other observations or prior information dominate the regression (detect using Cook's D, leverage, DFBETAS, or dimensionless scaled sensitivities).

    (c) There are errors in the conceptual representation of the system. For example, large differences between observed and advection-simulated ages can occur in flow fields that have a lot of abrupt divides in the direction of flow (such as in topographically controlled flow in shallow aquifers, or fault or structurally controlled flow systems), even though the main trends in the simulated age match the main trends in the observations. There can be a temptation to address the exceptionally poor-fitting observations in some way, such as adjusting the defined age, but these efforts are generally not useful to understanding system dynamics.

A more useful approach is to calculate leverage and Cook's D statistics for the observations. If the exceptional observations are not dominating estimated parameter values, it is best to leave them as part of the observation dataset. They may be important to future modeling efforts that represent the system in more detail, and will then be important to characterizing what are now unrepresented features in the model.

A small error in initial particle placement can result in large errors in simulated values. This issue is inherent to the approach and dependent on the particular hydrologic setting. Trying to sort out each error one-by-one can be problematic and introduces bias, as the analyst adjusts values to what they "should" be to reduce the error.

Resolution:

(a) It is important that the points where particles are introduced into the model grid be chosen with care. If particles are introduced on a high point in the water table from which flow diverges, and the high point moves around with differing parameter values, the resulting particle tracks can be very erratic. Likewise, if the entry point is close to a dominant groundwater sink, particles may track towards the sink and exit the system prematurely. Changing the entry point even slightly may result in more realistic particle tracks. Use a few trial runs to obtain entry locations that seem appropriate. Including additional particles is also recommended to improve simulation performance. If a problem is very sensitive to different, but equally likely, initial particle locations, discuss this in any description of the calibration effort.

(b) Carefully scrutinize the data from which observations and their weights are calculated. Observations that have large weighted residuals and scaled sensitivities will dominate the regression and should be examined first. Specifically, look at how the observations used with the particle tracking were obtained. Do they represent a composite surface sample from a well or spring, or are they less integrated depth-specific or depth-dependent samples? Alternatively, if the model grid-cell spacing is large, the water table has a steep gradient, and the observation locations are sparse across the region of steep gradient, model error may be a larger factor than anticipated. An observation that is causing a problem may be less precise than initially thought, so that the statistic from which the weight is calculated should be increased. Alternatively, the observation may be so inaccurate that omission from the parameter estimation needs to be considered. Any omissions need to be reported and justified in the model report.

(c) Any of the assumptions on which the conceptual model is based may be in error. Commonly, aquifer heterogeneity is oversimplified or incorrectly defined, or specified boundary conditions are in error. Again, scrutinize the model setup and revise model input.

3. Problem: Complex particle tracks can cause the regression not to converge.

Discussion: Heterogeneity can produce convoluted particle tracks that result in problematic regression runs. Anderman and Hill (2001) discuss a situation in which simulated heterogeneity produced such an erratic path that the regression did not converge, although the path had been produced by the synthetic problem considered. These can violate the smooth-function (linearity) requirement for convergence of regression.

Resolution: If intermediate or additional advective-transport data are available, use these in the parameter estimation as well. Otherwise, consider omitting the observation temporarily. Try to include the observation later in the calibration when the model more accurately represents the actual system.

4. Problem: Unexpected parameter values are estimated when advective-transport observations are included in the parameter estimation.

Discussion: Often, in finding the best fit to the data, the nonlinear regression will estimate parameter values that are unexpected. Re-examine the range of reasonable values and determine whether the unexpected values are unreasonable.

Resolution: If the parameter values are truly unreasonable, see Problem 5.

5. Problem: Unreasonable parameter values are estimated when advective-transport observations are included in the parameter estimation. This problem is also discussed by Hill (1998, p. 13) and Hill and Tiedeman (2007, p. 140-142), and is mentioned here because of its prevalence and importance.

Discussion: The estimation of unreasonable parameter values by regression provides information about likely model accuracy, and data accuracy and sufficiency. When advective-transport observations are included in the regression, they may provide information that is different than other types of data and can help identify model error. It is important during model development to consider unrealistic parameter estimates carefully and whether the groups of observations being used completely constrain the estimation process. Uncertainty in effective porosity may lead to uncertainty in parameter estimates, or parameter estimates that appear unreasonable. Simulated observations involving age are affected by the porosity values; if porosity is also embedded in the simulation of the storativity of the aquifer units, that are also being constrained by other groups of observations (such as water levels and water-level differences), then the storage estimates will also be affected by porosity values.

Resolution: Parameter estimation merely provides the best fit to the available data with a given model setup. For unreasonable parameter values, first consider the confidence interval on the estimate. If the confidence interval includes reasonable values, the unreasonable parameter value is NOT a clear indicator of model error. In this circumstance, consider adding constraints to the parameter value or additional types of observations that will contribute to constraining the estimation, and rerunning the regression. If the confidence interval does not include reasonable values, a problem is strongly indicated. Scrutinize the sources of available data, including information, such as in what season the data were observed, to determine (1) whether the data are correctly interpreted, (2) that the statistic used to calculate the weight correctly represents likely data error, and (3) that the simulated processes are adequate for the situation reflected in the data. If unreasonable parameter values still cannot be explained, reconsider the conceptual model and revise the model accordingly. For example, for porosity parameters, the use of strict upper and lower limits can help prevent estimation of unrealistic values that could affect the use of porosity for estimation of travel times, age dates, or storativity.

Linkage Between Programs

The performance of MODPATH-OBS is illustrated in figures 11A and B. For each forward run, MODPATH-OBS performs a sequence of steps to get the control information and MODPATH information, associates the particles to observation and source zones, and generates the global observations file for each type of user-specified observation category.

MODPATH-OBS

The flow chart in figure 11 illustrates the performance of MODPATH-OBS, including reading in the results from MODPATH-LGR. For each model grid, MODPATH-OBS retrieves model related data from the discretization file and particle data from the endpoint file (global endpoint file for MODPATH-LGR). Based on the user-specified observations, MODPATH-OBS develops the associations between source and observation locations for each observation. Then MODPATH-OBS calculates the simulated equivalents of the observations and writes output files for each observation type. MODPATH-OBS also produces instruction files for use with PEST or UCODE_2005, if specified by the user.

When simulated equivalents cannot be calculated, as when particles used to simulate time-of-travel observations do not reach the intended destination, it can sometimes be useful to consider an alternate simulated value. This can be accomplished using the program Sim_Adjust (Poeter and Hill, 2008). This capability can be especially important when using model analysis programs; Sim_Adjust is very general and can be used with UCODE, PEST, and other similar programs.

MODFLOW2000/2005 or MODFLOW-LGR

No changes were required in MODFLOW (versions 2000, 2005) or MODFLOW-LGR to accommodate MODPATH-OBS. MODPATH-OBS uses the discretization file(s) from the MODFLOW input directly; there is no other interaction between any version of MODFLOW and MODPATH-OBS. MODFLOW-LGR differs from MODFLOW-2005 only as needed to locally refine model grids, as described in Mehl and Hill (2005, 2007). Input files developed for MODFLOW-2005 can be used with MODFLOW-LGR and with MODPATH-OBS without modification when there is no local grid refinement.

MODPATH or MODPATH-LGR

MODPATH-OBS will work with the standard version of MODPATH or with MODPATH-LGR, but the latter enables more capabilities. In addition to changes to MODPATH to track particles within locally refined grids, as documented by Dickinson and others (2011), MODPATH-LGR includes the following to:

1. The global-endpoint output file now supplies global coordinates in all directions, including the vertical dimension. Previously, only vertical coordinates relative to the bottom of model layers were provided.

2. Multiple release times are allowed for backward as well as forward tracking.

3. Three integer identifiers and an optional character label are now provided for each particle. These labels are needed to identify which particles are used for each observation. The new information is listed in columns to the right of the previously existing columns, so the file should be compatible with existing software.

4. Global output files are now created for the combined output of multiple nested models.

Limitations

MODPATH-OBS is compatible with most capabilities of MODFLOW-LGR and MODPATH-LGR. An online guide and selected package incompatibilities are provided for MODFLOW-LGR (version 2) at http://water.usgs.gov/nrp/gwsoftware/modflow2000/MFDOC/index.html. Here we cite a few additional limitations that are relevant to MODPATH as follows:

1. MODPATH—Does not simulate the movement of particles through hydrologic features simulated using the following MODFLOW capabilities:

    a. Streamflow with the Streamflow-Routing (SFR) Package (Niswonger and Prudic, 2005) or the Streamflow Routing (SWR) Process (Hughes and others, 2012),

    b. Multi-Node Well (MNW) Package (Halford and Hanson, 2002; Konikow and others, 2009),

    c. Farm Process portion of MF-FMP (Schmid and others, 2006, 2009),

    d. PRMS portion of GSFLOW (Markstom and others, 2008),

    e. Conduit-Flow Package (CFP) (Shoemaker and others, 2007),

    f. Unsaturated Zone Flow (UZF) Package (Niswonger and Prudic, 2006), and

    g. Variably Saturated Zone Flow (VSF) Package (Thoms and others, 2006).

2. MODPATH-OBS—Proximities are currently calculated as simple rectilinear Euclidean distances. Alternatives not supported in the present version include distances traveled through selected parts of the subsurface. Such distances can be used, for example, to obtain an estimate of the time water spends in contact with selected rock types, which could relate to geochemical analyses.

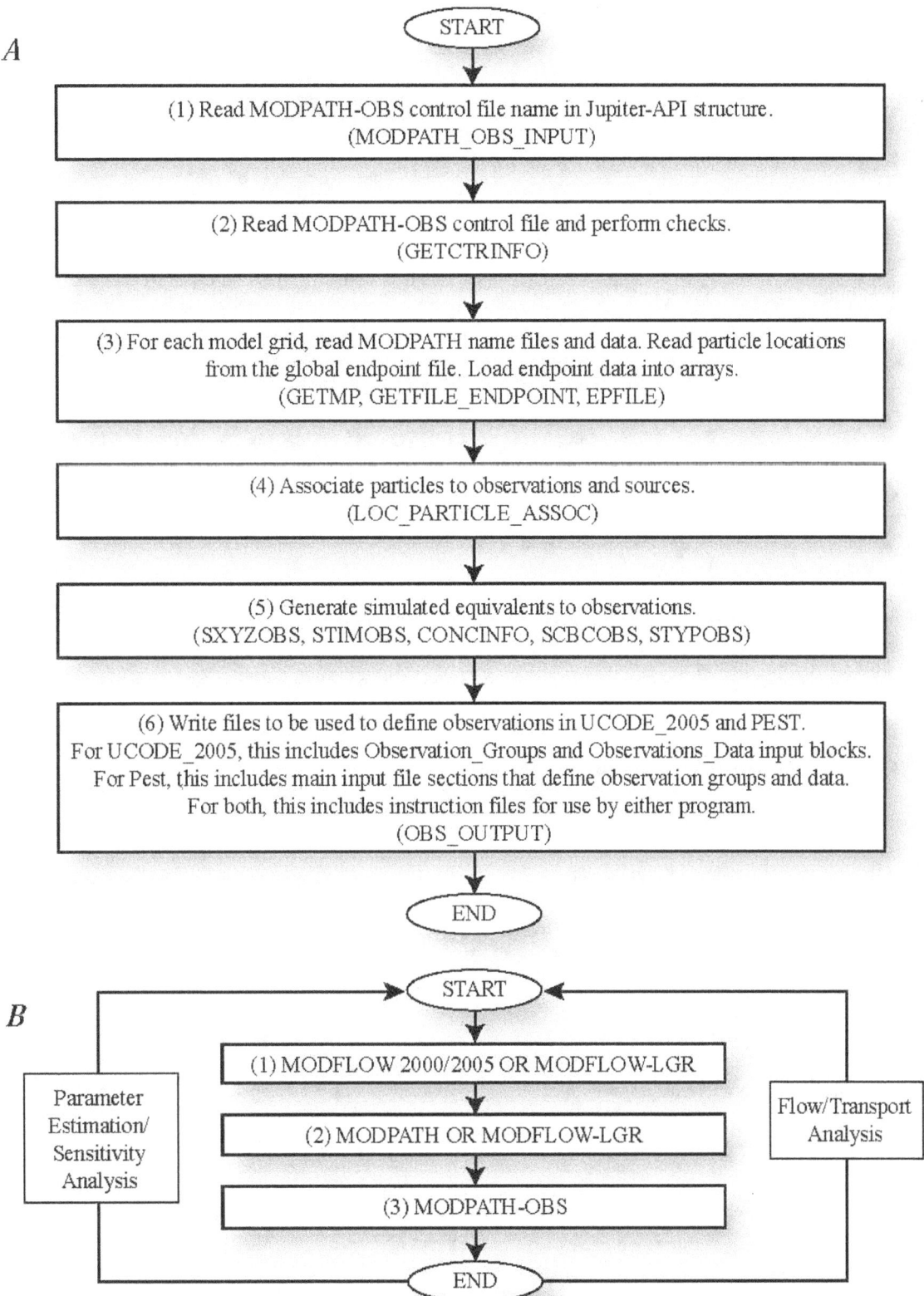

A

START

(1) Read MODPATH-OBS control file name in Jupiter-API structure.
(MODPATH_OBS_INPUT)

(2) Read MODPATH-OBS control file and perform checks.
(GETCTRINFO)

(3) For each model grid, read MODPATH name files and data. Read particle locations
from the global endpoint file. Load endpoint data into arrays.
(GETMP, GETFILE_ENDPOINT, EPFILE)

(4) Associate particles to observations and sources.
(LOC_PARTICLE_ASSOC)

(5) Generate simulated equivalents to observations.
(SXYZOBS, STIMOBS, CONCINFO, SCBCOBS, STYPOBS)

(6) Write files to be used to define observations in UCODE_2005 and PEST.
For UCODE_2005, this includes Observation_Groups and Observations_Data input blocks.
For Pest, this includes main input file sections that define observation groups and data.
For both, this includes instruction files for use by either program.
(OBS_OUTPUT)

END

B

START

Parameter
Estimation/
Sensitivity
Analysis

(1) MODFLOW 2000/2005 OR MODFLOW-LGR

(2) MODPATH OR MODFLOW-LGR

(3) MODPATH-OBS

Flow/Transport
Analysis

END

Figure 11. (*A*) Performance of MODPATH-OBS (program subroutine names are listed in parentheses), and (*B*) the sequence of programs needed for MODPATH-OBS For a batch file used with UCODE or PEST, if there are only porosity parameters, there is no need to repeat MODFLOW run for each estimation or sensitivity run.

3. MODPATH-OBS—Splitting and creation of particles along flow paths is not supported. Such methods were implemented in the particle-tracking analysis through multi-node wells for a study in Nebraska (Clark and others, 2008), but splitting and creating particles are not included in this version of MODPATH-OBS.

4. MODPATH-OBS—The particle tracking only represents advective transport within the saturated groundwater system, and can only represent observations derived from concentrations of a conservative constant-density geochemical constituent for which corrections have been made for the effects of mechanical dispersion and molecular diffusion. Obtaining accurate measures of advective transport in field situations can be difficult; the use of such measures in MODPATH-OBS may over or underestimate the actual attributes of transport. See the section of this report entitled "Relating simulated advective transport to field conditions" for additional information on this important aspect of using advective-transport observations and predictions. For example, Kauffman and Chapelle (2010) included longitudinal dispersion and had degradation rates that varied based on zones of the aquifer through which particles were traveling. McMahon and others (2008) allowed for land use change over time, whereas in the current MODPATH-OBS, there is no allowance for the source areas to change over time.

Hypothetical Example Problem

The example model demonstrates selected features of observations within the context of forward modeling, sensitivity analysis, parameter estimation, and uncertainty evaluation. The particle-tracking observations are defined for various source-location and observation-location associations that are then used for the estimation of transmissivity and porosity for five hydraulic property blocks.

Forward Modeling with MODPATH-OBS

The hypothetical problem represents a 400-m thick regional system with two pumping wells (fig. 12). Files for simulations with flow fields that are steady-state and transient-state are distributed with the code; results presented in this report are from the version with the transient-state flow field. The system is represented as confined for all simulations. The steady-state flow-field model is run with particles tracked backwards for an elapsed time of 50,000 years. The transient-state flow-field simulation initially reproduces the 50,000-year steady-state simulation and then creates 110 years of transient-state flow caused by changing recharge rates in

three areas of the model. Particles are tracked backwards in time starting at the end of the simulation, so they are affected first by the transient flow field with pumpage and then by the steady-state flow field without pumpage.

The distribution of hydraulic properties (fig. 12) is based on a laboratory experiment described in Garcia (1995) and Mapa and others, (1994), as discussed by Mehl and Hill (2002). These true values of transmissivity vary over more than three orders of magnitude, and are represented by five zones (fig. 12). The hydraulic conductivities and porosities do not change with depth. The initial porosity values range from 0.11 to 0.4 (fig. 12). For the transient-state problem, the specific storage is calculated using the assumed skeletal elastic compressibility of 4.7×10^{-6} m^{-1}, a water compressibility of 1.0×10^{-6} m^{-1} and the porosity distribution $\theta(x,y,z)$ used with MODPATH-LGR (fig.12). The equation used is

$$S_s = [4.7 \times 10^{-6} + \theta(x, y, z) \times (1 \times 10^{-6})] \qquad (3)$$

This is implemented in MODFLOW by using the MULT package. Thus, porosity affects storativity and particle velocity, which, in turn, affect the simulated transient head and particle observations.

Boundary conditions of the steady-state model are as follows (fig. 12):

*   Constant head of 0.0 m on the east side of all model layers;

*   A river along the west side of layer 1 with a stage of 11.0 meters simulated by using the River Package;

*   No-flow boundaries at the north, south, west (beneath the river), and bottom of the flow system;

*   Defined recharge at the farm, trench, and pit with concentrations that vary over time of perchloroethylene (pce) at the trench and the pit, and chlorofluorocarbon-12 (cfc) at all three sites; and

*   Pumping rates of 0.022 and 0.02 m^3/s at wells 2 and 3, respectively, distributed vertically in the top 200 m of the system.

The example problem is synthetic. The "true" model has a uniformly fine model grid and 12 model layers. The "true" model provides the "observed" values that are used as targets for the refined-grid model.

The locally refined model has a parent grid that is coarser than the true model and two child models with grid spacing equal to that of the true model (fig. 12). The parent grid has 50 rows and 108 columns with row and column discretization of 9.25 m and 9.0 m, respectively. The child grids each have 100 rows and 154 columns with cell dimensions of 1.028 m and 1.0 m in the row and column directions, respectively. In the locally refined model, the

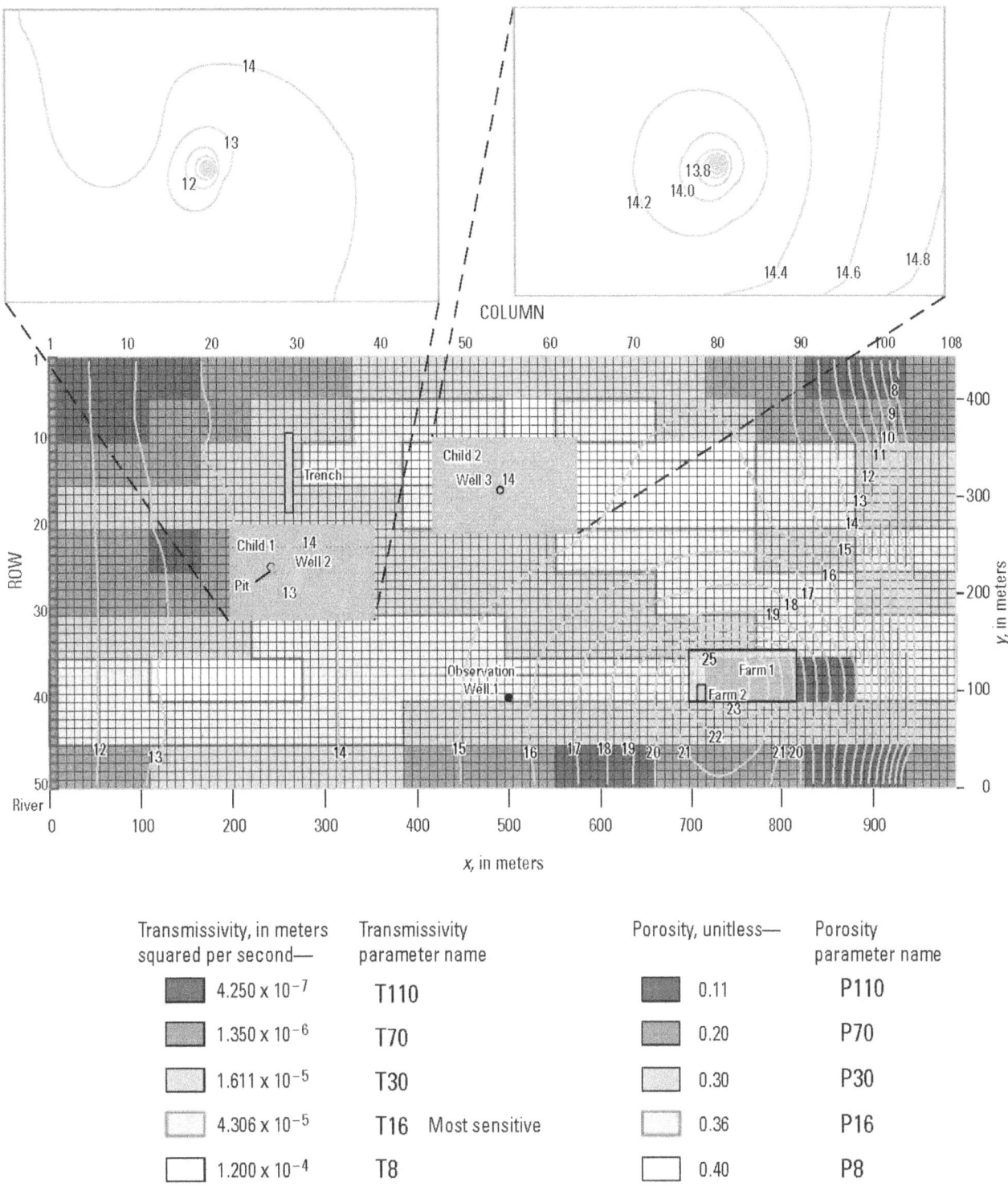

Figure 12. Hypothetical test problem showing steady-state head contours (meters) in model layer 1 using the locally refined model using true values of the parameters. (The heterogeneity pattern and parameters values are also shown. The two areas of local grid refinement are shown in green, labeled as child 1 and child 2. In these child models, layers one, two, and three of the parent model are simulated. Particle sources include the river along the left boundary, the trench, the farm, and the pit (the pit is in child model 1 to the left of the well). The four river wells are not used in the results presented in this report. Observation locations include wells 1, 2, and 3. Figure modified from Mehl and Hill, 2002.

resulting horizontal refinement ratio is 9:1 in both directions. The parent model has 4 model layers that are each 100 m in thickness. The two refined-grid child models each have 8 model layers. The child grids extend vertically 250 m from the top of parent layer one to the center of parent layer three using a 3:1 refinement; the layer thickness in the child models is 33.3 m for layers 1 to 7, and 33.3/2=16.65 for layer 8. Extending down to the center of a parent layer is consistent with the shared-node grid refinement method used. The locally refined model has the same distribution of hydraulic conductivity and porosity and the same boundary conditions as the "true" globally-refined grid model from which the observations are derived (fig. 12).

Steady-state heads, with pumping, produced for model layer 1 by the locally refined model is shown in figure 12. The figure clearly shows the high heads produced by recharge at the farm, the low heads produced by pumping at wells 2 and 3, and the heads at the boundary conditions on the left and right. The global flow budget produced by MODFLOW-LGR for the example model is shown in table 4.

Following an initial steady-state stress period, the 110 years of the transient flow field simulation are divided into four stress periods (2 through 5) as follows: 70, 20, 10, and 10 years. In the transient model the boundary conditions are as for the steady-state model except that defined inflows at the Farm, Trench and Pit are not constant. When flow occurs, the rates are the same as in the steady-state model, but flow only occurs during the following stress periods:

Farm: Stress periods 2 through 5 (110 year duration of inflow),

Trench: Stress periods 3 and 5 (20 years duration of inflow, 10 years no inflow, and 10 years duration of inflow), and

Pit: Stress periods 3 and 4 (30-year duration of inflow).

Particle paths calculated for the steady-state flow-field model are shown in figure 13A and 14. Concentrations assigned to inflows at the farm, trench, and pit for the MODPATH-OBS concentration observation type when either the steady-state or transient flow fields are used are shown in figure 13B.

Model Parameters

The ability to specify equations for parameters in MODPATH-OBS supports systematic analysis of basic system properties that affect model input parameters. In this example model, the defined parameters include transmissivities and porosities for each of the five zones shown in figure 12; thus, there are 10 defined parameters. There are no errors in how the parameters are defined – the distribution of the transmissivities and porosities is consistent with those in the true system. No

Table 4.   Inflows and outflows to and from the groundwater system for the steady-state example model simulated by using the locally refined model with true parameter values.

[True parameter values in cubic meters per second (m^3/s). Values are simulated by the parent model except as noted]

Inflow (m^3/s)		Outflow (m^3/s)	
River	0.0000	River	0.0117
Farm[1]	0.1481	Well2 (Child1)	0.0022
Trench[1]	0.0017	Well3 (Child2)	0.0200
Pit[1] (Child1)	0.0003	Constant head	0.1161
Total	0.1508 (0.1501)	Total	0.1508 (0.1500)

[1]Recharge values are derived from the following calculations:
Farm: 78 cells × 83.25m^2/cell × 2.28×10^{-5}m/s=0.1481 m^3/s.
Trench: 9 cells × 83.25m^2/cell × 2.28×10^{-6}m/s=0.0017 m^3/s.
Pit: 1 cell × 1.028m^2/cell × 2.8×10^{-4}m/s=0.0003 m^3/s.

prior information on parameters is included in regression runs, but constraints are imposed on parameter values. In this example model, porosity is represented by the Darcian velocities of the particles in MODPATH-LGR and in the definition of storativity built through the Multiplier package of MODFLOW. Because storativity is related to porosity which is, in turn, fundamentally related to grain-size distribution and dewatering and when used in this dual capacity as a model input parameter, it can strongly influencing particle-tracking results.

Observations and Observation Weighting

The observations that are used as surrogates for field observations are created using the observations from the globally refined-mesh model as the "true" observations (fig. 12). No noise is added except due to rounding the numbers to values that are consistent with the typical accuracy of field measurements.

The 109 "true" observations include 15 hydraulic heads and 94 transport observations that include all four types documented in this report: proximity, time-of-travel, concentration, and source water percentages (tables 5 and 6). There are five head observations at each of the three wells. The observed head values are composite heads from all layers in the true refined model derived from the vertical averaging in the HOB package using proportions that are the same for each model layer. The head observations are weighted based on whether they are located in the parent grid, with its large grid size, or a child grid. At well 1, which is in the parent model, the weight is calculated using a standard deviation of 0.5 m. At wells 2 and 3, which are in the child models, the weight is calculated using a standard deviation of 0.3 m.

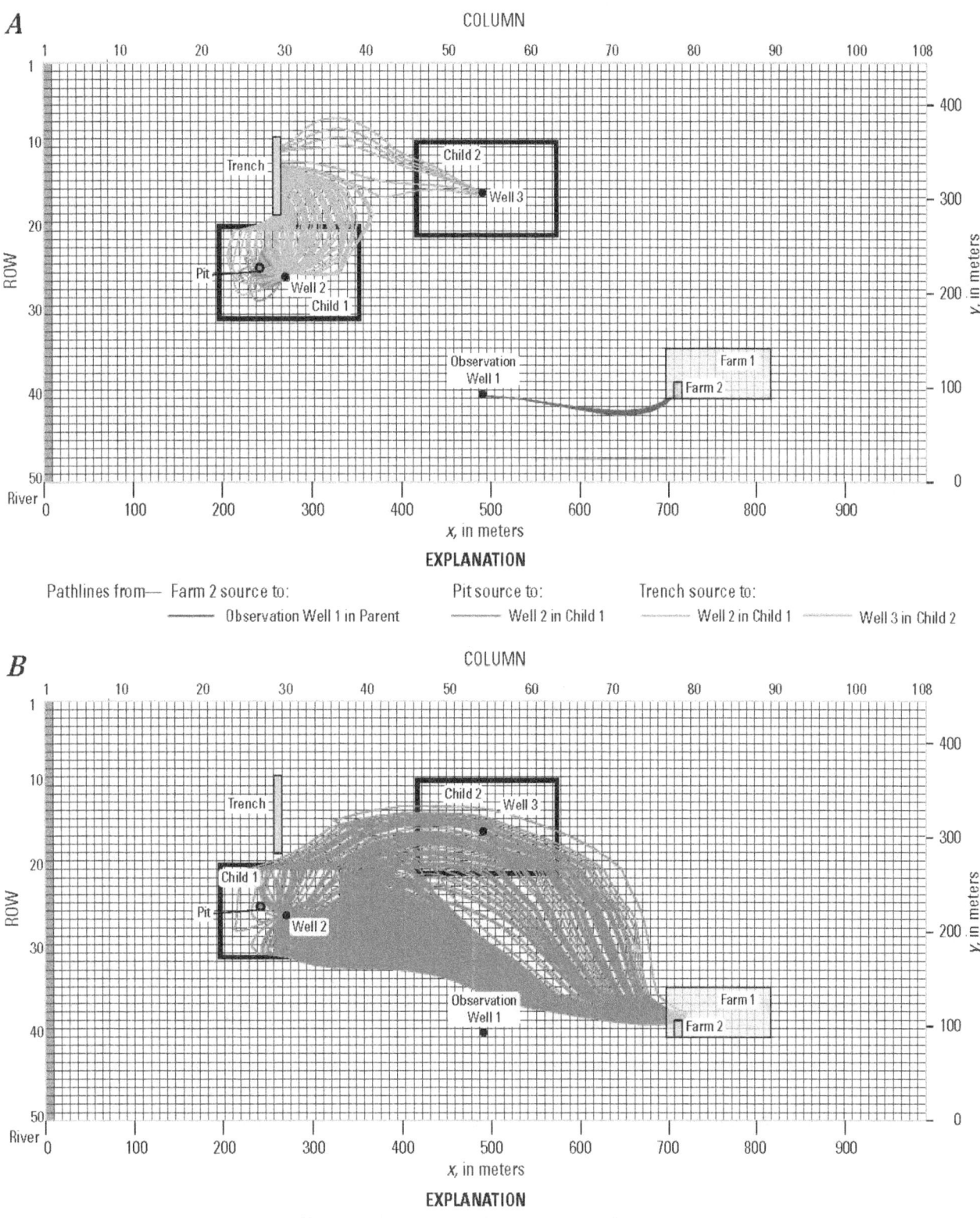

EXPLANATION

Pathlines from— Farm 2 source to: Pit source to: Trench source to:

———— Observation Well 1 in Parent   ———— Well 2 in Child 1   ———— Well 2 in Child 1   ·········· Well 3 in Child 2

EXPLANATION

Pathlines from Farm 1 source to: ———— Well 2 in Child 1

Figure 13.   Plan view of simulated pathlines from the hypothetical test problem with the steady-state flow field and related capture zones and endpoint particles for each pumping well for (*A*) sources to Farm 2; (*B*) sources to Farm 1 from well 2 (child model 2); (*C*) sources to Farm 1 from well 3 (child model 2); (*D*) concentrations histories used by MODPATH-OBS for flow from the farm, trench, and pit; and (*E*) concentration histories for the transient flow field simulation, recharge at the farm is absent for the predevelopment steady-state starting stress period, and constant thereafter; recharge at the trench vary in time as shown.

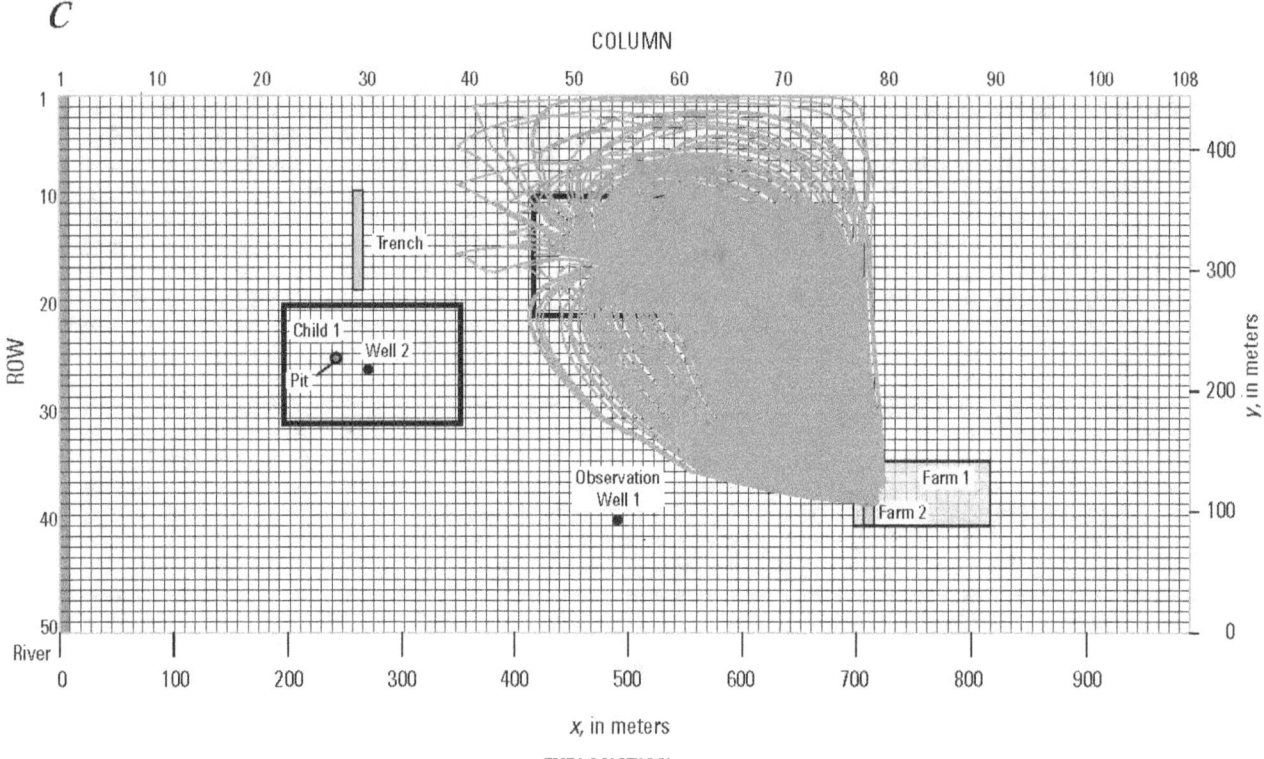

C

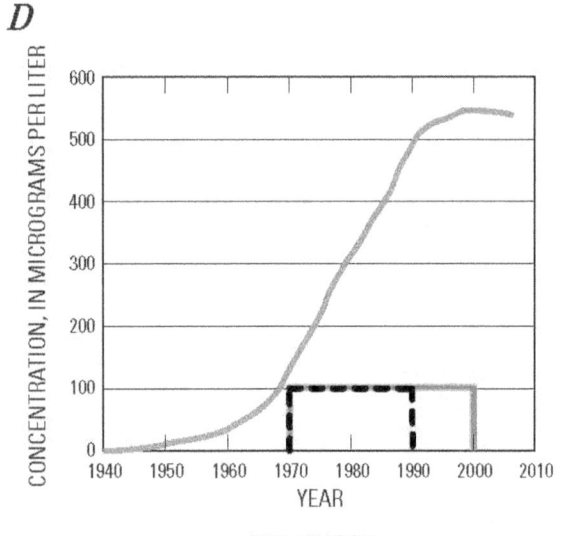

D

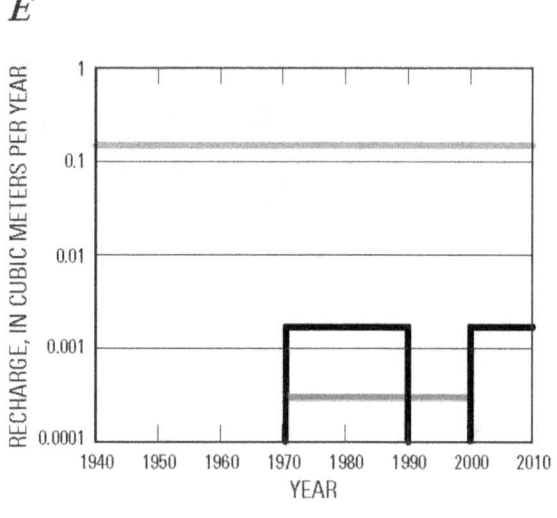

E

EXPLANATION

Pathlines from—

Farm 1 source to:

——— Well 3 in Child 2

EXPLANATION

——— CFC (farm, trench, and pit)

——— PCE (pit)

▬▬▪ PCE (trench)

EXPLANATION

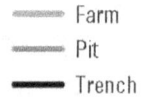

▬▬▬ Farm

▬▬▬ Pit

▬▬▬ Trench

Figure 13.—Continued

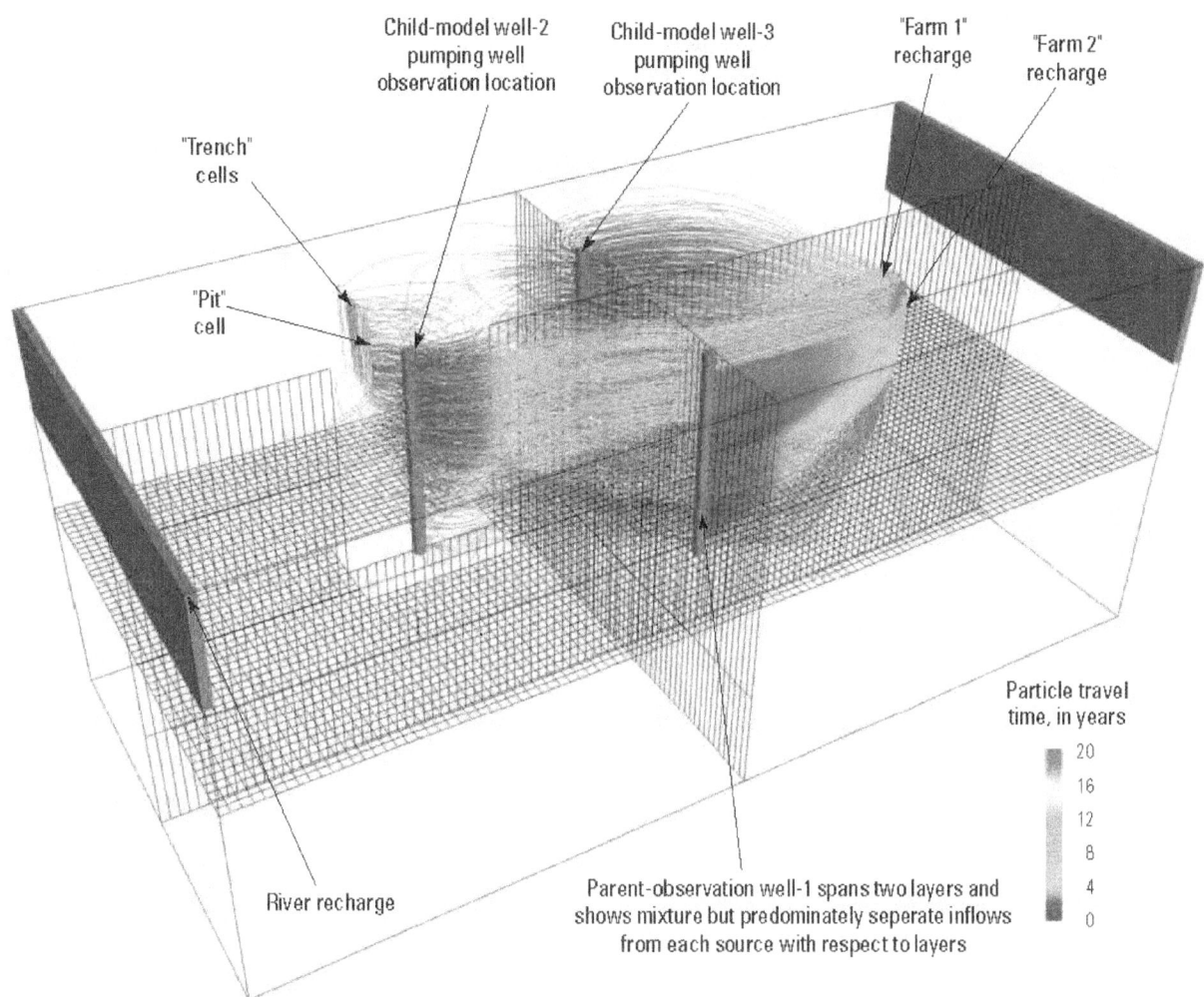

Figure 14.   Backward tracked particles for the hypothetical test problem in 2010 under steady-state flow conditions. Colors for the particle pathlines are in years.

The transport observations represent samples from wells 1, 2, and 3. These observations are designed to characterize transport from four source zones to these wells. The four source zones occur in the upper model layer and include the river, the farm, and the trench in the parent model, and the pit source in child model 1. Pathlines between selected components of the system are shown in figures 13 and 14, respectively. To produce the particle paths needed to obtain the "observations," the true globally refined model was used to simulate subsurface flow and track 1,018 particles over a 100 year period from 1910 to 2010. The particles were tracked backward from the observation locations to the source locations. The percent water from four sources at the three wells defined for the example problem shown in table 5 illustrates how mixtures are formed from multiple sources.

The observations used for the example problem that are simulated using MODPATH-OBS are listed in table 6. There are 6 proximity observations, 13 time-of-travel, 50 concentration, and 25 source-water type observations. Attributes are listed in table 6, including the source and the observation location. In forward particle tracking, the particles would be tracked from the source to the observation location, and the SampleTime would be set to the time the particle arrives at the observation location. In backward tracking, as used for all observations in this work, the particles are tracked from the observation location to the source. In this case, SampleTime is the time the backward-tracked particle was released from the observation location.

The combination of observation types used in this example helps to reduce parameter correlation between transmissivity and porosity values and thus supports unique estimation of more of the defined parameters. Selected sensitivity-analysis, parameter estimation, and uncertainty-evaluation results from the transient-state flow simulations are presented to demonstrate the utility of this method.

The weighting for all transport observations are calculated using a standard deviation of 4.47 derived from the simulation of the fine-mesh model. Table 6 shows that simulated values range from 0 to 423 for cnc_cfc3. Such a wide range of values suggests that a variable value of the standard deviation is likely to better represent the observations errors, which is common for lognormally distributed observations. However, a constant value is used in this work.

Table 5.   Percent of water from four sources that reaches three wells in 2010.

[Wells 2 and 3 are pumped, well 1 is not. The Zone_ID is defined in the Source_Zones_### input blocks, where ### is replaced by the model name (here parent, child1, or child2), (A) Simulated using the globally refined transient flow field model with true parameter values. Thus, these are the values used as observations in model calibration; (B) Simulated using the locally refined model with transient-state flow field and true parameter values; (C) Observation names associated with each of the values.]

(A)

Zone_ID		Well 1	Well 2	Well 3
1	Pit	–	1	0
3	Trench	–	6	1
4	River	–	76	1
5	Farm	71	11	92
2	Farm2	–	–	–

(B)

Zone_ID		Well 1	Well 2	Well 3
1	Pit	–	1	0
3	Trench	–	6	0
4	River	–	81	1
5	Farm	64	10	99
2	Farm2	–	–	–

(C)

Zone_ID		Well 1	Well 2	Well 3
1	Pit	–	tyW2Pit2010	typ_pit3
3	Trench	–	tyW2Tr2010	typtrench3
4	River	–	typ_river2	typ_river3
5	Farm	typfarmto1	typ_farm2	typ_farm3
2	Farm2	–	–	–

Table 6. Observation attributes for the hypothetical example for which the simulated equivalent is produced using particle tracking for the transient flow-field model.

[In addition to the listed 94 observations, 15 head observations are defined at three locations, five times each. The observations listed in this table are in file obsdata.txt, part of which is listed in appendix B. See figure 12 for location of sources and observation locations For most entries of OBS_NAME, the first three letters define the observation type (xyz, proximity; tim, time-of-travel; cnc, concentration; typ, souyes, normalityrce water type), single digit numbers (2 or 3) identify the associated well. For Source_ID, all means the pit, trench, farm and river; pce means the source of pce (perchloroethylene), which include the pit and trench; cfc means the source of cfc (chlorofluorocarbon), which includes the pit, trench, farm and farm2. For Comp_Type, Min is minimum, Med is median, Max is maximum, PctGE100 and PctLT100 define the value reported from the exceedence plot as being the percent greater or equal to 100 or less than 100, respectively]

Observation identifier (OBS_Name)[1]	Name of observation-location location (OBSLoc_ID)	Source (Source_ID)	Sampling-time, year is listed[2] (SampleTime)	Observed value[3] (ObsValue)	Comparison component (Comp_Type)
PROXIMITY, in meters from observation cell {OBSType=proximity; there are 5 observations} [4]					
xyzobs1-3	well1_obs	farm2	2010	[5]0, 0, 0	x, y, z-distance (Min)
xyzobs4-5	well2-3_obs	farm2	2010	[5]0, 0	Total-distance (Min)
TIME-OF-TRAVEL, in elapsed years relative to observation {OBSType=time; there are 14 observations}[6]					
timmedpit2	well2_obs	pit	2010	23.38	(Med)
timmedpit2t	well2_obs	pit	2010	1.37	(Med)—Log[7]
timminpit2 a-f	well2_obs	pit	2005–2010	7.4–23.4	(Min)
timmaxpit2	well2_obs	pit	2010	23.4	(Max)
timminall3	well3_obs	all	2010	3.22	(Min)
timminriver3	well3_obs	river	2010	127.68	(Min)
timminfarm3	well3_obs	farm	2010	3.22	(Min)
timge1003	well3_obs	all	2010	1.04	(PctGe100)
timlt1003	well3_obs	all	2010	98.9	(PctLT100)
CONCENTRATIONS , in parts per million {OBSType=conc; there are 50 observations.}[6, 8]					
cnc_cfc1	well1_obs	cfc	2010	372.5	Conc
cnc_pce1	well1_obs	pce	2010	0. 0	Conc
cnc_cfc2a-c	well2_obs	cfc	2008-2010	243.6–258.3	Conc
w2pce1970-2010	well2_obs	pce	1970-2010	2.0–8.1	Conc
w2pceExc1	well2_obs	pce	1	67.7	Exceedance (Exc1[8])
w2pceExc2	well2_obs	pce	2	50.	Exceedance (Exc2[8])
cnc_cfc3	well3_obs	cfc	2010	422.7	Conc
cnc_pce3	well3_obs	pce	2010	0.57	Conc
SOURCE WATER TYPE, in percent {OBSType=source; there are 25 observations}					
typfarmto1	well1_obs	farm	2010	70.59	Percent
typ_farm2	well2_obs	farm	2000	40.4	Percent
typ_river2	well2_obs	river	2010	10.9	Percent
typ_pit3	well3_obs	pit	2010	0.0	Percent
typtrench3	well3_obs	trench	2010	0.78	Percent
typ_farm3	well3_obs	farm	2010	92.4	Percent
typ_river3	well3_obs	river	2010	0.78	Percent
tyW2Pit1970-2010	well2_obs	pit	1970–2010	0.0–4.7	Percent
tyW2Tr1970-2010	well2_obs	trench	1970–2010	0.0–5.7	Percent

[1] The variable name shown in parenthesis is defined and explained in appendix A. Model names listed here with a dash indicate a sequence of observations. For example, xyzobs1-3 identifies observations xyzobs1, xyzobs2, and xyzobs3.

[2] For forward particle tracking, elapsed time is added to the reference time. For backward particle tracking, elapsed time is subtracted from the reference time.

[3] Weighting for observations is calculated using the following statistics: for heads, standard deviation equals 1.0 or 2.0; for all other observations, standard deviation equals 4.47. For proximity observations, the location for the particle is the location identified by the source. The proximity observed value used in regression and listed in the UCODE-2005 input files is zero because the goal is for the relative distance (i.e. proximity) from the location to be zero.

[4]No observed distance is specified for the proximity distance observations. The final location of the observation is derived from the particle location from the endpoint file and is compared against the location of the observation to estimate a proximity distance.

[5] Zero means that the particle(s) reached the intended destination or position in the x, y, or z direction.

[6] Both time and particle-concentration observations have the option of log transformation. The values used in this example are not transformed. PctGe100 is the percentage of particle travel times greater than or equal to 100 years. PctLT100 is the percentage of particle travel times less than or equal to 100 years. Min is minimum value, Max is maximum value, Med is the median value.

[7] The observation is log-transformed using the Keyword Transform in the MODPATH-OBS Observation_Data input block.

[8] Particle-concentration observations are made as the simple difference between the total particles at the specified concentration for that point in the time history of that contaminant.

[9] Defined using ExDefID in the Exceedance_Definition Input Block.

Parameter Prior Information and Reasonable Ranges

It is common for data and knowledge about sediment and geologic processes and system geometry to provide information about transmissivity (which is the product of hydraulic conductivity and thickness) and storage properties in groundwater models. This information can be used as part of regression in three ways: to identify unreasonable parameter values that can be a key indicator of model inadequacy (Hill and Tiedeman, 2007, p. 140-142), (2) in the regression as prior information, which is very useful in obtaining good regression results, and (3) in the regression to constrain the values of parameter estimates (Hill and Tiedeman, 2007, Guideline 5). The problem here uses this information in the first two ways. The UCODE input and out related to prior information are listed in table 7. Table 7 also includes the calculation that commonly is used to link the UCODE input for prior information in the Linear_Prior_Information input block and the input for reasonable upper and lower parameter values in the Parameter_Data input block.

Table 7. UCODE input and output related to parameter prior information and reasonable upper and lower limits, both of which would be supported by information independent of the observations used in the regression.

[UCODE, UCODE_2005 is used. Std dev, standard deviation. UCODE input is highlighted in orange. The UCODE input for prior information is listed in the Linear_Prior_Information input block. The reasonable upper and lower parameter values are listed in the Parameter_Data input block. These values do not constrain the regression–estimated values are compared with them to make the modeler aware of unreasonable parameter value estimates UCODE output is highlighted in gray. When parameters are log-transformed as in this problem, UCODE presents log10 results as much as possible.]

Parameter name	UCODE input to define prior information		Listed in UCODE Output		Associated confidence interval limits on $\log_{10}(b)$ $(A \pm 2B)$		Exp10 of confidence interval limits— used as UCODE input reasonable upper and lower values		True parameter value
	(A) Prior value	(B) Std dev $\log_{10}(b)$	$\log_{10}b$	Weight for $\log_e(b)$	Lower	Upper	Lower	Upper	
Groupname T									
T110Prior	1×10^{-7}	0.5	−7.00	0.75	−8.00	−6.00	1×10^{-8}	1×10^{-6}	4.25×10^{-7}
T70Prior	5×10^{-7}	0.5	−6.30	0.75	−7.30	−5.30	5×10^{-8}	5×10^{-6}	1.30×10^{-6}
T30Prior	6×10^{-6}	0.5	−5.22	0.75	−6.22	−4.22	6×10^{-7}	6×10^{-5}	1.61×10^{-5}
T16Prior	5×10^{-5}	0.5	−4.30	0.75	−5.30	−3.30	5×10^{-6}	5×10^{-4}	4.31×10^{-5}
T08Prior	1×10^{-4}	0.5	−4.00	0.75	−5.00	−3.00	1×10^{-5}	1×10^{-3}	1.20×10^{-4}
Groupname P									
P110Prior	0.20	0.13	−0.70	4.72	−0.96	−0.44	0.11	0.36	0.11
P70Prior	0.25	0.13	−0.60	4.72	−0.86	−0.34	0.14	0.45	0.20
P30Prior	0.25	0.13	−0.60	4.72	−0.86	−0.34	0.14	0.45	0.30
P16Prior	0.25	0.13	−0.60	4.72	−0.86	−0.34	0.14	0.45	0.36
P08Prior	0.25	0.13	−0.60	4.72	−0.86	−0.34	0.14	0.45	0.40

Parameter Estimation, Model Fit, and Sensitivity Analysis

This section provides some selected model calibration results from the hypothetical problem with the transient flow field. These results demonstrate the use of the MODPATH-OBS observation types in model development. The results from parameter analyses are mostly presented through graphs for the three sets of parameters (table 8) with the parameter values are listed in table 9. Results from the true parameter values cannot be shown because this is a synthetic test case. The misfit at the true parameter values is due to using a locally refined grid instead of the globally refined grid used to generate the values used as observations and because simulated values are truncated to a number of significant digits that are more representative of field observations. No noise was added to create the observations. Notes in table 8 provide some guidance for the reader and additional commentary is provided in the following sections.

Model Fit

Overall measures of model fit presented at the top of table 8 are the sum of squared weighted residuals (SSWR) and standard error. The model fit for the calibrated model (SSWR=96.2) is a bit better than for the true model because of the fitting process of the regression. There are parameter adjustments that can make up for the model error produced by using a locally refined grid instead of a globally refined grid, and the regression finds those adjustments. The progression of both parameter values and overall measures of model fit are shown in figure 15. Figure 15A shows how some parameter values jump around. This performance made it difficult to achieve the preferred convergence criteria of parameter values changes by 1% of the value from one iteration to the next. In figure 15B, the number of observations is listed because some of the observation types in MODPATH-OBS cannot be calculated. Thus, with MODPATH-OBS, this value can vary over the course of the parameter estimation procedure.

Model fit can also be evaluated by comparing graphs of simulated and observed values over time. Figure 16 shows four sets of concentrations: Observed, simulated, starting, and true. The latter is known for this synthetic test case. The observed and simulated values are plotted in the first set of model fit graphs in table 8, along with similar values for the rest of the observations. Weighting the simulated and observed values allows a more meaningful comparison, and is shown in the second row of table 8. Finally, at each time in figure 16, the observed minus simulated values (the residuals) can be calculated. These differences are weighted and used to evaluate model fit in the third set of model fit graphs in table 8.

In all graphs of model fit, observed values are plotted on the horizontal axis except for the graph with weighted values. Using observed values makes it easier to see what happens to the fit of specific observations for the three sets of parameter values considered. Often simulated values are used on the horizontal axis, but these values are different for the different sets of parameter values and make it impossible to follow the fit of a given observation.

Sensitivity Analysis

The sensitivity analysis results show what parameters are informed by the observations and what observations are most important to the parameters.

Parameter importance to observations is measured using bar charts of composite scaled sensitivities (CSS) together with parameter correlation coefficients (PCC) and bar charts of the parameter t-statistic (ln(b)/SD). These are shown in the first two rows of the sensitivity analysis part of table 8. For both, the height of the bars measure the total information provided by each parameter. Differences occur primarily because the parameter-t-statistic includes the effects of PCC, while CSS does not so that PCC is reported separately. The results for four of the transmissivity parameters are consistent, but for all three parameter sets the t-statistic suggests that T110 is less identifiable than the CSS evaluation indicates. Both methods suggest that P08 and P110 are likely to be difficult to estimate. The marginal locations of T110 and P110 relative to the observations (fig. 12) are consistent with the lack of importance of these parameters and the likely difficulty in estimating these parameter values. The other three porosity parameters are rated as more identifiable by CSS than the t-statistic, and the PCC statistic calculated for the true parameter values suggests the contribution of parameter value interdependence to the different results.

For CSS, the colors within the bars show how each observation type contributes to this measure. The least important observations for this problem are the proximity observations. These observations are most useful when the flow filed is grossly in error, and that situation does not occur for the present problem. The other MODPATH-OBS observation types and heads are consistently important, though their contribution changes somewhat as the parameter values and resulting simulated flow field changes.

Equations for CSS, PCC, and the parameter standard deviation used to calculate the t-statistic are described by Hill and Tiedeman (2007, p. 48-54 and 126-127).

When parameters cannot be estimated using the available observations, direct information on the parameters if often used to achieve a tractable problem. Here this is accomplished by adding prior information to all of the parameters, as described in the section entitled "Parameter Prior Information and Reasonable Ranges". Here the prior information added is thought to accurately reflect knowledge about the parameter values. If larger weights (smaller variances and standard deviations) are needed to obtain a tractable regression, care needs to be taken because uncertainty evaluations will express greater confidence that is warranted (see Hill and Tiedeman, 2007, p. 304-305).

The importance of each observation to the set of estimated parameters is indicated in table 8 using Cook's D and leverage. The equations for these measures can be found in many works, including Hill and Tiedeman (2007, p. 59, 135).

Cook's D is a scaled measure of the change in the set of parameter values that would occur if the observation is omitted, and accounts for the value of each observed value.

Leverage depends only on the type, time, and location of the observations, and not the observed value. Leverage equal to 1.0 identifies observations that dominate one or more of the estimated parameters. This does not mean the other observations are not important at all—it means that more than one observation contains redundant and largely consistent information about the parameter values so that the observations do not individually control one or more parameter values.

For both Cook's D and leverage, results for the starting and estimated parameter values are calculated with the prior information defined. However, Cook's D equals zero if the residual equals zero, as it does for prior information when evaluated at the starting values, and this is why no prior information shows up as important in the Cook's D graph for the starting parameter values. The results for the true parameter values do not have prior information defined.

Results of the Cook's D and leverage evaluation in table 8 suggest that some observations and prior information are consistently dominant. For example, timminriver3 is consistently important, as are some of the concentrations plotted in figure 16. The important concentrations tend to occur when the concentration changes and (or), for Cook's D, when the observed minus simulated (residual) values are large.

The individual parameters that are most affected by the observations identified as being important can be evaluated using the DFBETAS statistic and dimensionless scaled sensitivities (DSS). Graphs that include the 15 observations for which these values are the largest are shown in table 8.

Table 8. Model fit, sensitivity analysis, and parameter uncertainty graphs for the hypothetical example showing results from parameter values used to start the regression, estimated by regression, and true values[1] known for this synthetic test case.

[The model has a transient flow field and a locally refined grid. On log scales, 0.0 values are plotted as 0.1]

Results for starting and true parameter values[1]	Results for estimated parameter values[1,2]
Model Fit	
SOSWR=5606 for starting values and 125 for true values (observations only)	OSWR=92 for observations, 96 for when prior information is added
Standard error=7.2 and 1.12	Standard error=0.94

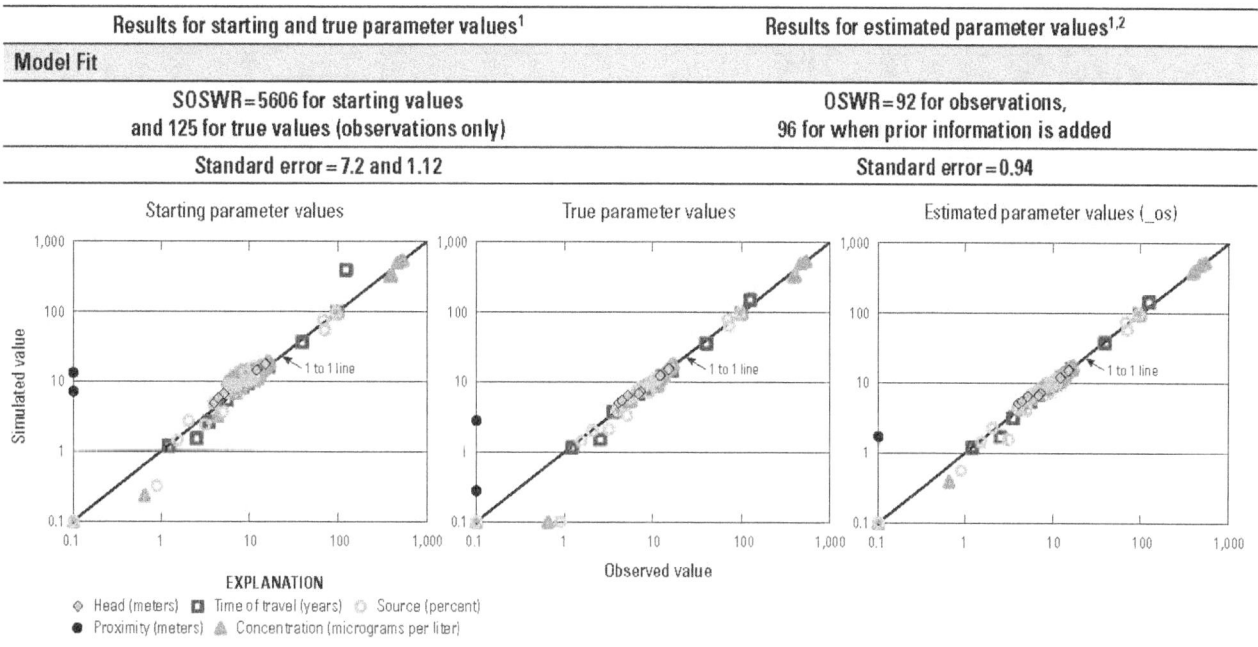

Weighting reveals differences that fall within expected levels of uncertainty. Notice the proximity observations.

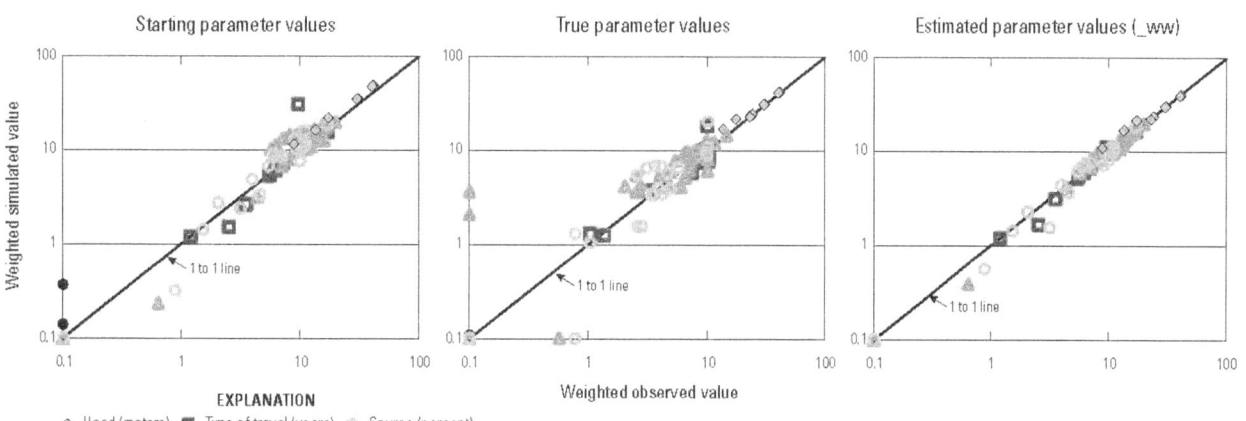

Table 8.   Model fit, sensitivity analysis, and parameter uncertainty graphs for the hypothetical example showing results from parameter values used to start the regression, estimated by regression, and true values[1] known for this synthetic test case.—Continued

[The model has a transient flow field and a locally refined grid. On log scales, 0.0 values are plotted as 0.1.

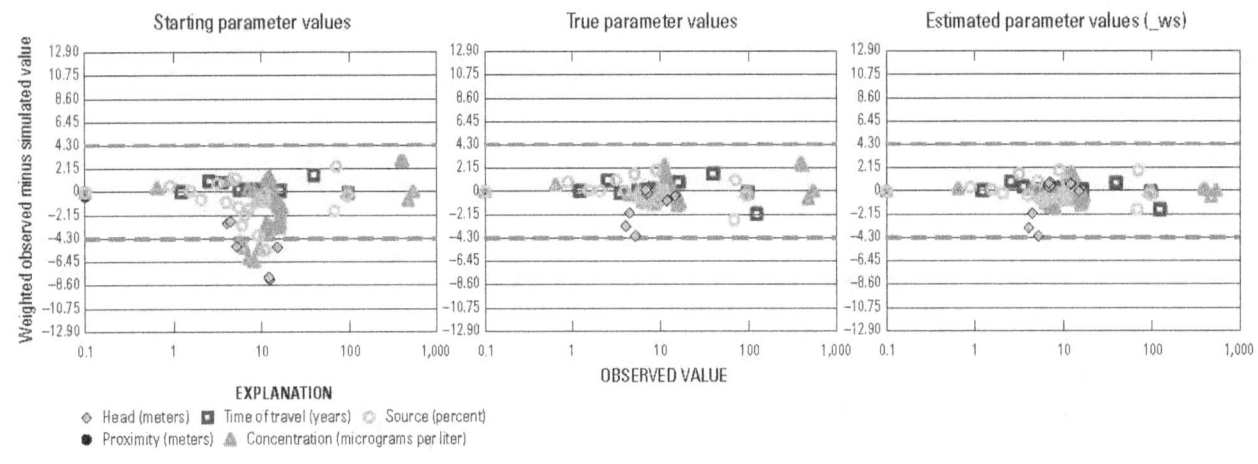

EXPLANATION

◇ Head (meters)   ▣ Time of travel (years)   ◎ Source (percent)
● Proximity (meters)   ▲ Concentration (micrograms per liter)

Vertical axis increments are the standard error of the weighted residuals for estimated values. If the weighted residuals were normally distributed, on average 5 percent of the values would fall outside the heavy dashed lines.

Table 8. Model fit, sensitivity analysis, and parameter uncertainty graphs for the hypothetical example showing results from parameter values used to start the regression, estimated by regression, and true values[1] known for this synthetic test case.—Continued

[The model has a transient flow field and a locally refined grid. On log scales, 0.0 values are plotted as 0.1.

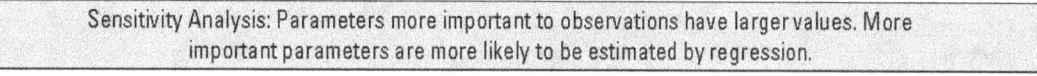

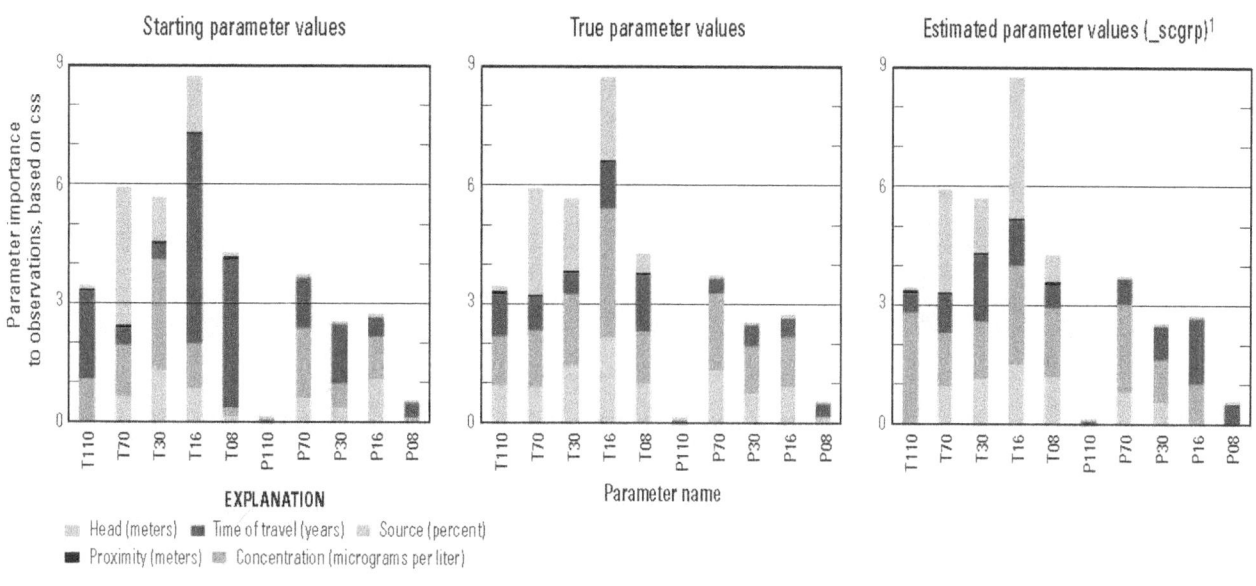

Parameter interdependence as measured by parameter correlation coefficient (PCC) absolute values larger than 0.85. For the starting and estimated values prior information results in no high PCC values):

none   P110:P08 PCC = -0.92   none

This measure of parameter importance reflects the combined effects of CSS and PCC.

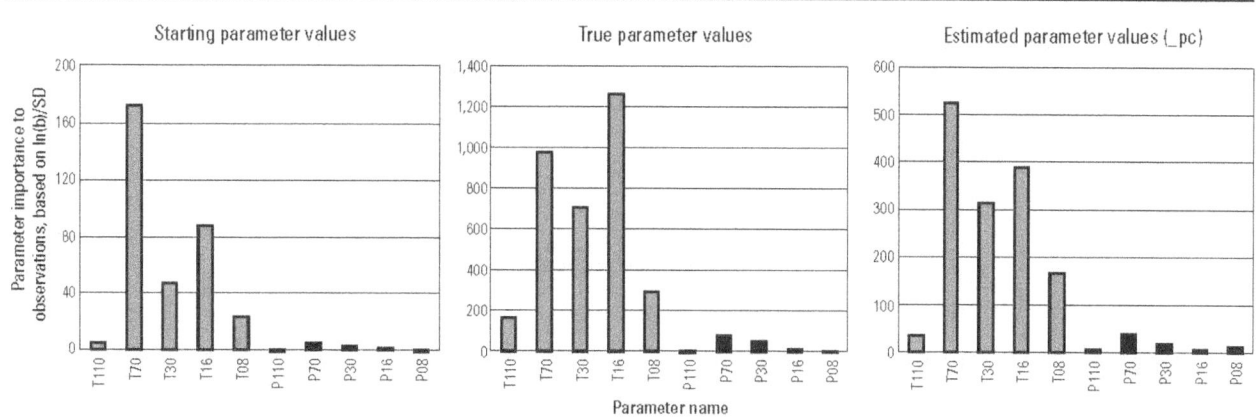

Table 8. Model fit, sensitivity analysis, and parameter uncertainty graphs for the hypothetical example showing results from parameter values used to start the regression, estimated by regression, and true values[1] known for this synthetic test case.—Continued

[The model has a transient flow field and a locally refined grid. On log scales, 0.0 values are plotted as 0.1.

Sensitivity Analysis: Observations and prior information more important to parameters have larger values. More important observations are more likely to merit additional investigation to reduce or characterize errors in the value or simulated equivalent. (No prior information is defined for the results for the true parameter values)

Cook's D identifies actually important observations (accounts for the observed value). The critical value is 0.04 and is shown by a dotted line when the scale allows. Larger values indicate large changes in the estimated parameters would result if the observations were omitted.

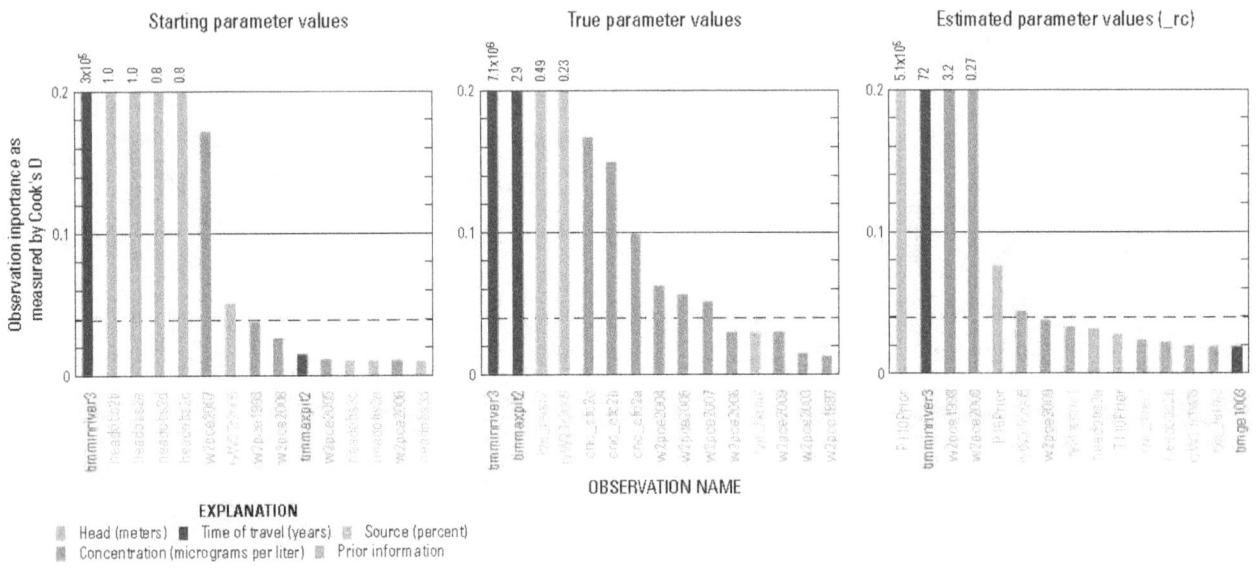

Leverage identifies potentially important observations (does not account for the observed value). Leverage = 1 identifies observations that completely dominate the value of at least one estimated parameter value. Progressively lesser values indicate that parameter values are estimated using more observations.

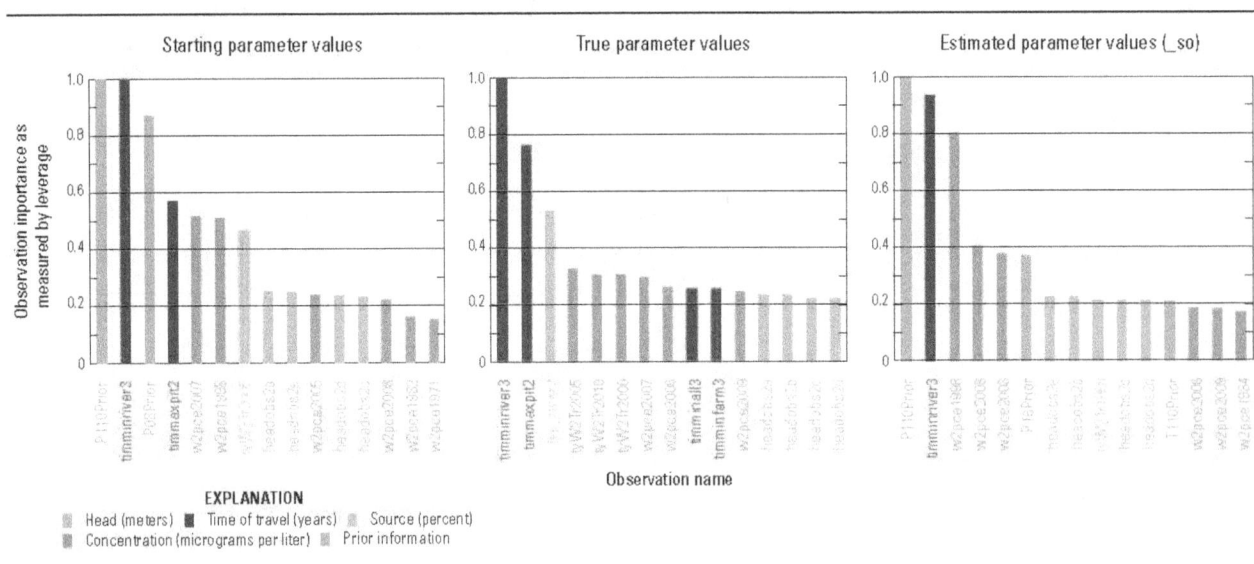

Table 8. Model fit, sensitivity analysis, and parameter uncertainty graphs for the hypothetical example showing results from parameter values used to start the regression, estimated by regression, and true values[1] known for this synthetic test case.—Continued

[The model has a transient flow field and a locally refined grid. On log scales, 0.0 values are plotted as 0.1.

Sensitivity Analysis: Observations more important to parameters have larger values. More important observations are more likely to merit additional investigation.(No prior information is defined for the results for the true parameter values)

DFBETAS indicates how much each parameter would change if the observation were omitted from the regression. It integrates the effects of sensitivity as measured by DSS and parameter interdependence as measured by PCC.

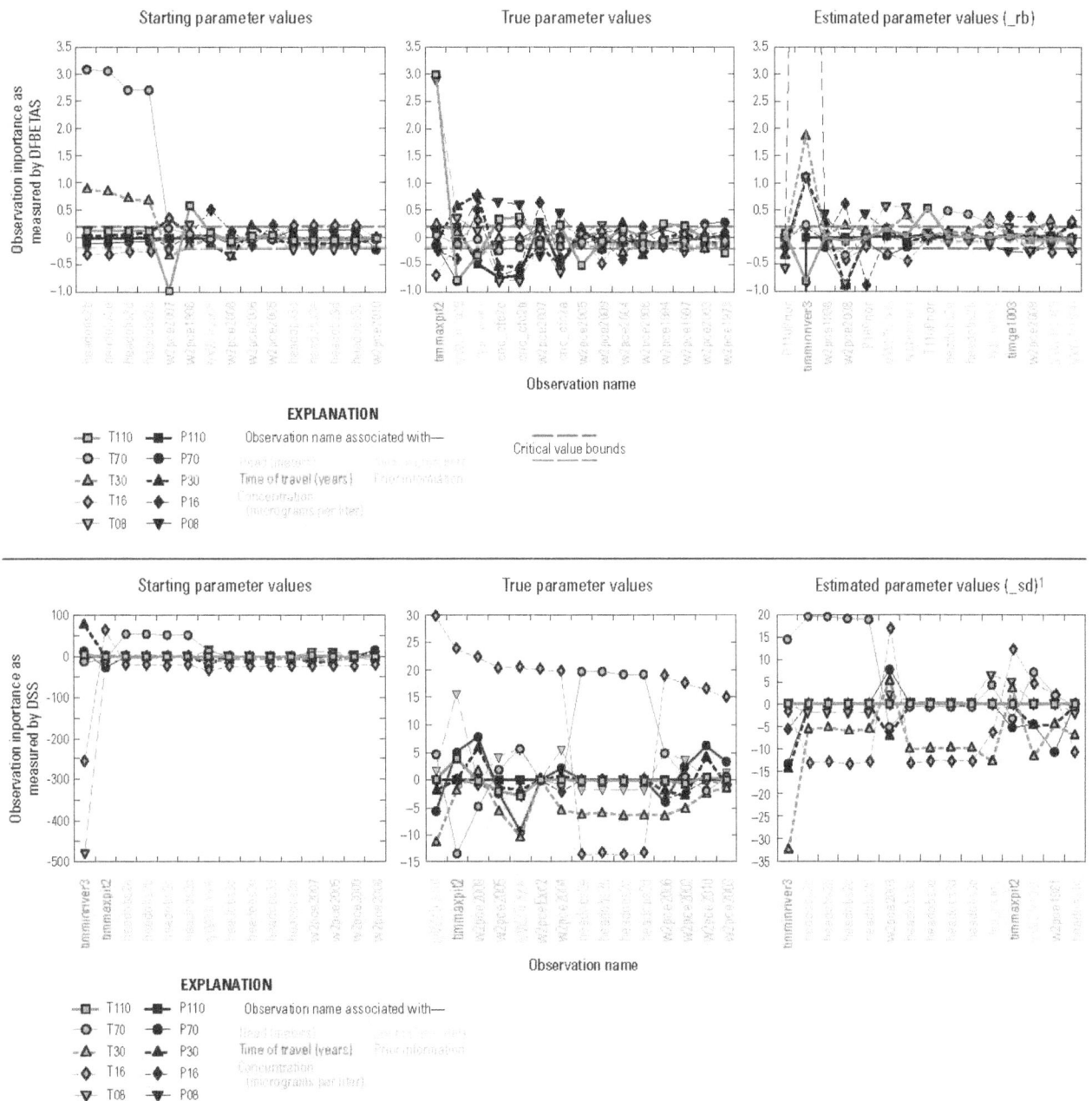

[1] The results for the true parameters do not include any defined prior information. Thus, all results reflect only observations. CSS and DSS account only for observations for all three parameter sets.

[2] The extension for the UCODE_2005 data exchange file used is in parentheses. For the sensitivity analysis, names of data-exchange files produced by UCODE_2005 start with "_s", and names of data-exchange files produced the postprocessor RESIDUAL_ANALYSIS start with "_r".

Table 9.   The starting, estimated, and true parameter values
and associated values of the sum of squared, weighted residuals
objective function.

[Parameter units: T in m²/d and P are dimensionless]

Parameter name	Parameter values		
	Starting	True[1]	Estimated
T110	1×10^{-7}	4.250×10^{-7}	2.510×10^{-7}
T70	5×10^{-7}	1.350×10^{-6}	1.354×10^{-6}
T30	6×10^{-6}	1.611×10^{-5}	1.550×10^{-5}
T16	5×10^{-5}	4.306×10^{-5}	4.993×10^{-5}
T08	1×10^{-4}	1.200×10^{-4}	7.501×10^{-5}
P110	0.20	0.11	0.20
P70	0.25	0.20	0.23
P30	0.25	0.30	0.27
P16	0.25	0.36	0.36
P08	0.25	0.40	0.27
Model fit statistics			
Sum of squared, weighted residuals	6,506	125	96
Standard error	7.2	1.1	0.94
Confidence interval on the standard error	6.3:8.3	0.99:1.3	0.83:1.1

[1] The only source of error for this model is the conversion from globally
refined to locally refined grids. The misfit displayed here is explained
completely by that difference.

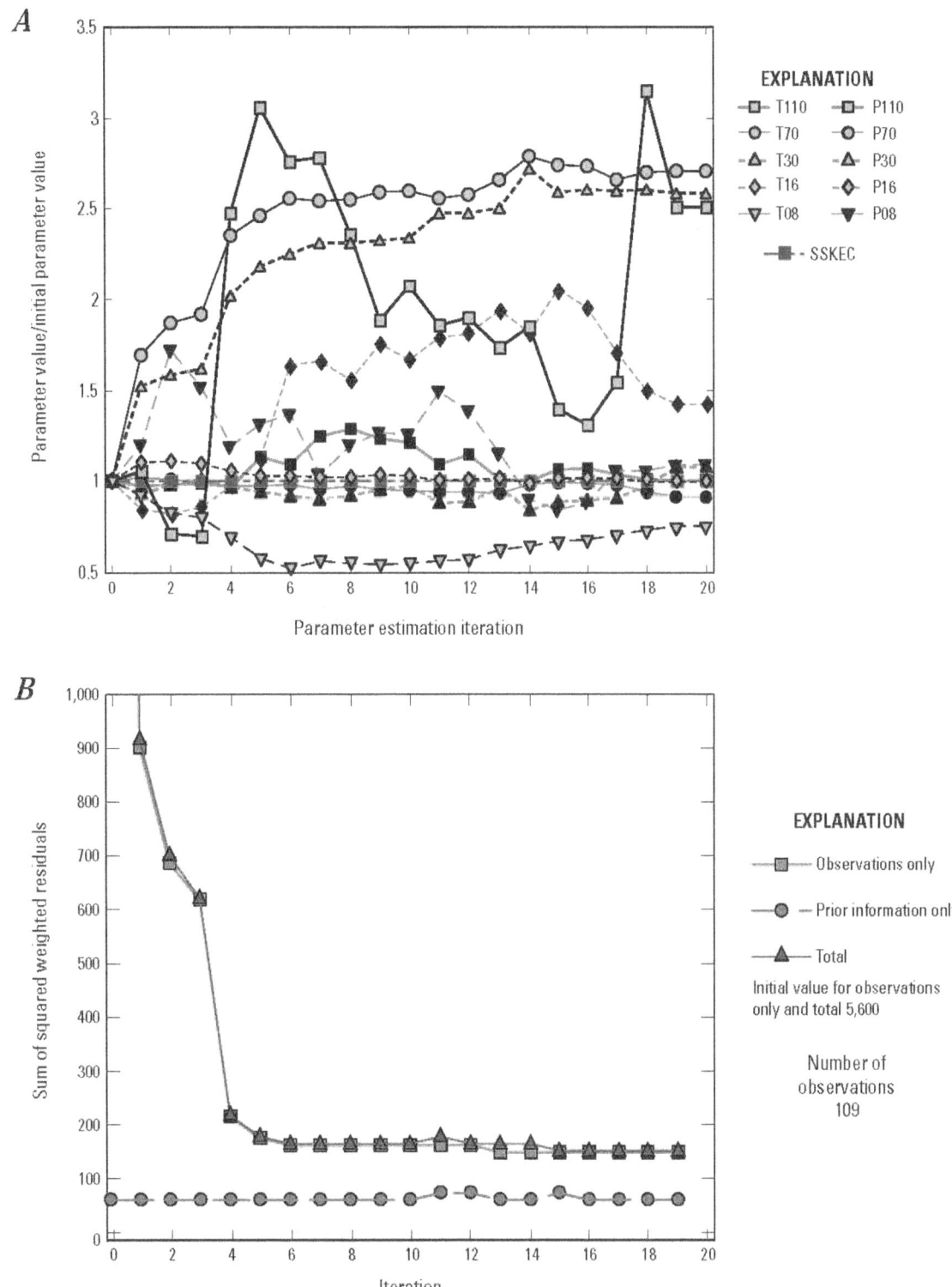

Figure 15. Performance of parameter estimation. (*A*) Changes in the parameter value during the parameter estimation iterations. Convergence occurred when no parameter value changed by more than 7% from one iteration to the next. (*B*) Change in model fit during the parameter estimation iterations. At the starting parameter values (iteration=0) the value is 6,506.

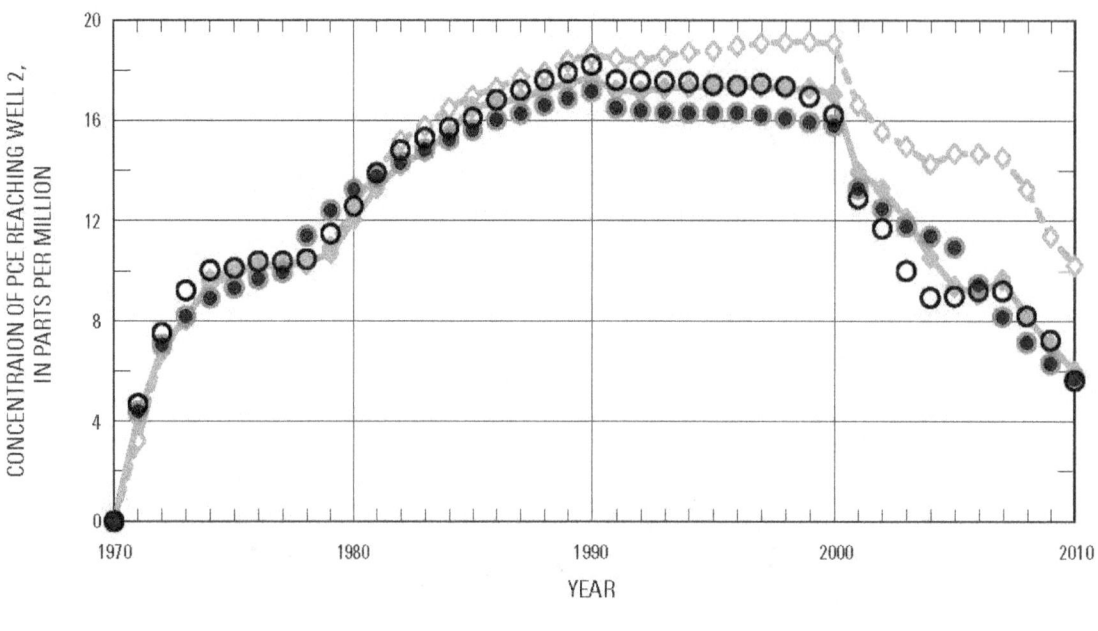

EXPLANATION

⬤   Observed (True parameter values with globally refined grid)

○   True parameter values with locally refined grid (Use of a locally
     refined grid is a model inadequacy)

⬦   Starting parameter values (Before regression)

⬦   Estimated parameter values (After regression; parameter values
     try to accommodate model inadequacy)

Figure 16.   Concentration of PCE reaching well 2, in ppm for four circumstances: the true parameters values used in a globally refined model (used to generate observations and three sets of parameters using the locally refined grid (true, staring, and estimated).These correspond to observations w2pce**, where ** is replaced by year 1970 to 2010. High values of leverage and DSS tend to occur where concentrations change rapidly in time. Cook's D and DFBETAS are highest where both the concentrations change rapidly in time and the difference between simulated and observed values is large.

Summary

This work describes a program developed to calculate simulated equivalents to transport observations derived from simulated particle tracking with or without field measurements for a variety of common field situations or hydrologic settings. The method can be used for transient and steady-state flow simulations. A hypothetical model is presented to demonstrate the method and to illustrate the application and considerations for use.

The model involved can include embedded locally refined grids to provide additional detail as needed. For example, additional detail is often useful around features such as landfills, disposal sites, springs, and wells. The program, MODPATH-OBS, works with MODPATH-LGR or MODPATH-5/MODPATH-6 and either MODFLOW-LGR, MODFLOW-200 or MODFLOW-2005 to calculate simulated equivalents to transport observations from tracked particles. The particles can traverse a parent model and up to nine embedded child models. Another potential future use for embedded models with particle tracking includes the analysis of weak sources and sinks. For example, the analysis of spring flow at Devil's Hole in the Armargosa Desert will utilize a subregional child model within the Death Valley regional model for more detailed simulation of spring discharge subject to nearby agricultural pumping. Also the Southern Amargosa Embedded Model (SAMM) project is using an LGR subregional model to evaluate the effects of pumpage within the Death Valley regional model (Wayne Belcher, USGS, written commun., 2012).

The inclusion of observations derived from particle tracking in the parameter estimation, sensitivity, and uncertainty analysis indicate that:

1. The use of observations that reflect the dynamics of paths traveled through a system provides considerable additional constraints on the estimation of parameters.

2. Locally refined grids allow greater resolution of features and dynamics likely to affect such paths.

3. Locally refined models enable use of data that are local in scale that generally are not viable in regional-scale models. For example, the water arriving at a well from different sources can be more accurately simulated so that the resulting mixture can be more reliably compared with measured values.

4. Using locally refined grids in inverse modeling allows the regional and local phenomena to be reconciled simultaneously and reduces the aliasing of the parameter values caused by inadequate resolution of local features. That is, the estimated parameter values are not compromised by the need to make up for poorly simulated dynamics as values are determined that produce the best match to simulated heads and flows.

5. Child models from the calibrated locally refined model can be used independently to obtain quick execution times when evaluating local scenarios. This can be useful in evaluating local contaminant transport. Local pumping and water-supply issues can also be evaluated, as long as the interfacing boundary conditions are not affected by the simulated conditions. This can be evaluated using the BFH-package capability of MODFLOW-LGR. The advantage is that the model calibration includes responses to local stresses, and all models are constrained by the observations derived from particle tracking throughout the locally refined grid.

6. The use of observations derived from particle tracking for parameter estimation of fully coupled regional-embedded models can provide accurate results for both flows and hydraulic heads, but can result in difficult numerical convergence in highly heterogeneous systems. However, the level of complexity of nested models and observation types can be adjusted to the scale of the problem, hydrologic setting, and analysis issues.

The current version of MODPATH-OBS focuses on relating observations to simulated equivalents in the context of particle tracking across the multiple grids that comprise a locally refined model. Future uses could extend this methodology to linkage to other models and additional processes and packages within MODFLOW. Many of the limitations stated above could be overcome by allowing particle tracking throughout more of the hydrologic cycle. Additional features, such as the use of probability density functions and additional forms of observations (for example, probability of arrival times), could be implemented as another way to compare time or concentration data types. In addition, by using PEST, UCODE, or another code, sensitivity analysis, uncertainty evaluation, and data-needs assessment can be further investigated. Particle tracking allows such evaluations to be accomplished quickly, perhaps as a preliminary step to using a much more computationally demanding advective-dispersive transport simulation.

The ability to specify equations for parameters in MODPATH-OBS supports systematic analysis of basic system properties that affect model input parameters. For example, porosity is fundamentally related to grain-size distribution and compaction caused by dewatering and related land-subsidence and is a model input parameter strongly influencing particle-tracking results. In the example problem, the use of particle-tracking observations facilitates the application of parameter estimation through derived variables such as the influence of porosity on aquifer storativity as well as its influence on the Darcian flow velocities through aquifers.

References Cited

Alley, W.M., 1993, Regional ground water quality: New York, Van Nostrand Reinhold, 635 p.

Anderman, E.R. and Hill, M.C., 2001, MODFLOW-2000, the U.S. Geological Survey modular groundwater model—Documentation of the advective-transport observation (ADV2) package, version 2: U.S. Geological Survey Open-File Report 01–54, 69 p.

Anderman, E.R., Hill, M.C., and Poeter, E.P., 1996, Two-dimensional advective transport in groundwater flow parameter estimation: Ground Water, v. 34, no. 6, p. 1001–1009.

Aster, R.C., Borchers, Brian, and Thurber, C.H., 2012, Parameter estimation and inverse problems (2d ed.): Waltham, Mass., Academic Press, 376 p.

Ballard, Sanford, 1996, The in situ variable flow sensor: A ground-water flow velocity meter, Ground Water, v. 34, no. 2, p. 231–240.

Banta, E.R., Poeter, E.P., Doherty, J.E., and Hill, M.C., 2006, JUPITER: Joint Universal Parameter IdenTification and Evaluation of Reliability—An application programming interface (API) for model analysis: U.S. Geological Survey Techniques and Methods, book 6, chap. E1, 268 p.

Barlebo, H.C., Hill, M.C., Rosbjerg, Dan, and Jensen, K.H., 1998, Concentration data and dimensionality in groundwater models—Evaluation using inverse modeling: Nordic Hydrology, v. 29, p. 149–178.

Bethke, C.M., and Johnson, T.M., 2008, Groundwater age and groundwater age dating: Annual Review of Earth and Planetary Sciences, v. 36, p. 121–152.

Burow, K.R., Jurgens, B.C., Kauffman, L.J., Phillips, S.P., Dalgish, B.A., and Shelton, J.L., 2008, Simulations of groundwater flow and particle pathline analysis in the zone of contribution of a public-supply well in Modesto, eastern San Joaquin Valley, California: U.S. Geological Survey Scientific Investigations Report 2008–5035, 41 p., http://pubs.usgs.gov/sir/2008/5035/.

Campana, M.E., and Simpson, E.S., 1984, Groundwater residence times and recharge rates using a discrete state compartment model and C-14 data: Journal of Hydrology, v. 72, p. 171–185.

Clark, B.R., Landon, M.K., Kauffman, L.J., and Hornberger, G.Z., 2008, Simulations of ground-water flow, transport, age, and particle tracking near York, Nebraska, for a Study of Transport of Anthropogenic and Natural Contaminants (TANC) to public-supply wells: U.S. Geological Survey Scientific Investigations Report 2007–5068, 49 p.

Clark, I.D., and Fritz, Peter, 1997, Environmental isotopes in hydrogeology: Boca Raton, Fla., Lewis Publishers, 328 p.

Clement, T.P., 2010, Complexities in Hindcasting Models—When should we say enough is enough?: Ground Water, v. 49, no. 5, p. 617–769.

Constantz, Jim, Cox, M.H., and Su, G.W., 2003, Comparison of heat and bromide as ground water tracers near streams: Ground Water, v. 41, no. 5, p. 647–656.

Cook, P.G., and Böhlke, J.K., 2000, Determining timescales for groundwater flow and solute transport, chap. 1 in Cook, P.G., and Herczeg, A.L., eds., Environmental tracers in subsurface hydrology: Boston, Mass., Kluwer Academic Publishers, p. 1–30.

Cook, P.G., and Herczeg, A. L., 2000, Environmental tracers in subsurface hydrology: Boston, Mass., Kluwer Academic Publishers, 529 p.

Crandall, C.A., Kauffman, L.J., Katz, B.G., Metz, P.A., McBride, W.S., and Berndt, M.P., 2008, Simulations of groundwater flow and particle tracking analysis in the area contributing recharge to a public-supply well near Tampa, Florida, 2002–05: U.S. Geological Survey Scientific Investigations Report 2008–5231, 53 p.

Deb, Kalyanmoy, 2001, Multiobjective optimization using evolutionary algorithms: New York, John Wiley & Sons, Inc., 518 p.

Dickinson, J.E., Hanson, R.T., Mehl, S.W., and Hill, M.C., 2011, MODPATH-LGR—Documentation of a computer program for particle tracking in shared-node locally refined grids using MODFLOW-LGR: U.S. Geological Survey Techniques and Methods book 6, chap. A38, 41 p., http://pubs.usgs.gov/tm/tm6a38/.

Dickinson, J.E., James, S.C., Mehl, S.W., Hill, M.C., Leake, S.A., Zyvoloski, G.A., Faunt, C.C., and Eddebbarh, A.A., 2007, New ghost-node method for linking different models with varied grid refinement and initial investigations of heterogeneity and nonmatched grids: Advances in Water Resources, v. 30, no. 8, p. 1722–1736.

Dixon, E.C., and Peterson, Kathleen, 2003, Utilization of a technical peer review to support the mission of the Nevada Test Site Community Advisory Board: WM'03 Conference, Tucson, Ariz., February 23–27, 2003, 12p., http://www.wmsym.org/abstracts/2003/pdfs/586.pdf.

Doherty, John, 2007, PEST Users Guide: Coriander, Australia, Watermark Computing, 122 p., http://pesthomepage.org/.

Domenico, P.A., and Schwartz, F.W., 1990, Physical and chemical hydrogeology: West Sussex, England, John Wiley & Sons Inc., 824 p.

Eberts, S.M., Böhlke, J.K., Kauffman, L.J., and Jurgens, B.C., 2011, Comparison of particle-tracking and lumped-parameter age-distribution models for evaluating vulnerability of production wells to contamination: Hydrogeology Journal, v. 20, no. 2, p. 263–282, http://dx.doi.org/ 10.1007/s10040-011-0810-6.

Foglia, Laura, Hill, M.C., Mehl, S.W., and Burlando, Paolo, 2009, Sensitivity analysis, calibration, and testing of a distributed hydrological model using error-based weighting and one objective function: Water Resources Research, v. 45, no. 6, 18 p., http://dx.doi.org/10.1029/2008WR007255.

Franke, O.L., Reilly, T.E., Pollock, D.W., and LaBaugh, J.W., 1998, Estimating areas contributing recharge to wells lessons from previous studies: U.S. Geological Survey Circular 1174, 14 p.

Garcia, J.E., 1995, An experimental investigation of upscaling of water flow and solute transport in saturated porous media: unpublished Master's Thesis, University of Colorado, Boulder, 135 p.

Graham, M.F., and Smart, G.T., 1980, Reservoir simulator employing a fine-grid model nested in a coarse-grid model: Annual Society of Petroleum Engineers Fall Technical Conference, 55th, Dallas, Tex., September, 21–24, 1980, Paper SPE 9372, 7 p.

Green, C.T., Puckett, L.J., Böhlke, J.K., Bekins, B.A., Phillips, S.P., Kauffman, L.J., Denver, J.M., and Johnson, H.M., 2008, Limited occurrence of denitrification in four shallow aquifers in agricultural areas of the United States: Journal of Environmental Quality, v. 37, no. 3, p. 994–1009.

Halford, K.J., and Hanson, R.T., 2002, User guide for the drawdown-limited, multi-node well (MNW) package for the U.S. Geological Survey's modular three-dimensional finite-difference ground-water flow model, versions MODFLOW-96 and MODFLOW-2000: U.S. Geological Survey Open-File Report 02–293, 33 p., http://pubs.usgs.gov/of/2002/ofr02293/text.pdf.

Hanson, R.T., Anderson, S.R., and Pool, D.R., 1990, Simulation of groundwater flow and potential land subsidence, Avra Valley, Arizona: U.S. Geological Survey Water-Resources Investigation Report 90–4178, 41 p., http://pubs.er.usgs.gov/usgspubs/wri/wri904178.

Hanson, R.T., 1996, Postaudit of head and transmissivity estimates and groundwater flow models of Avra Valley, Arizona: U.S. Geological Survey Water-Resources Investigation Report 96–4045, 84 p., http://pubs.er.usgs.gov/usgspubs/wri/wri964045.

Hanson, R.T., Wentworth, C.M., Newhouse, M.W., Williams, C.F., Noce, T. E., and Bennett, M.J., 2002, Santa Clara Valley Water District multiple-aquifer monitoring-well site at Coyote Creek Outdoor Classroom, San Jose, California: U.S. Geological Survey Open-File Report 02–369, 4 p., http://pubs.usgs.gov/of/2002/ofr02369/.

Harbaugh, A.W., 2005, MODFLOW-2005, The U.S. Geological Survey modular groundwater model—The groundwater flow process: U.S. Geological Survey Techniques and Methods, book 6, chap. A16, variously paged.

Hill, M.C., 1998, Methods and guidelines for effective model calibration: U.S. Geological Survey Water-Resources Investigations Report 98–4005, 90 p.

Hill, M.C., 2006, The practical use of simplicity in developing groundwater models: Ground Water, v. 44, no 6, p.775–781.

Hill, M.C., Banta, E.R., Harbaugh, A.W., and Anderman, E.R., 2000, MODFLOW-2000, the U.S. Geological Survey modular groundwater model—User guide to the observation, sensitivity, and parameter-estimation processes and three post-processing programs: U.S. Geological Survey Open-File Report 00–184, 210 p.

Hill, M.C., and Tiedeman, C.R., 2007, Effective groundwater model calibration: Hoboken, N.J., Wiley Interscience, John Wiley & Sons, 455 p.

Hunt, R.J., Coplen, T.B., Haas, N.L., Saad, D.A., and Borchardt, M.A., 2005, Investigating surface-water-well interaction using stable isotope ratios of water: Journal of Hydrology, v. 302, no. 3, p. 154–172.

Hughes, J.D., Langevin, C.D., Chartier, K.L., and White, J.T., 2012, Documentation of the Surface-Water Routing (SWR1) Process for modeling surface-water flow with the U.S. Geological Survey Modular Ground-Water Model (MODFLOW-2005): U.S. Geological Survey Techniques and Methods, book 6, chap. A40 (Ver. 1.0), 113 p.

International Atomic Energy Agency, 2006, Use of chlorofluorocarbons in hydrology—A guidebook: Vienna, Austria, International Atomic Energy Agency, 277 p., http://www.iaea.org.

Izbicki, J.A., Stamos, C.L., Nishikawa, Tracy, and Martin, Peter, 2004, Comparison of groundwater flow model particle tracking results and isotopic data in the Mojave River groundwater basin, southern California, USA: Journal of Hydrology, v. 292, no. 2, p. 30–47.

Izbicki, J.A., Christensen, A.H., and Hanson, R.T., 1999, U.S. Geological Survey combined well-bore flow and depth-dependent water sampler: U.S. Geological Survey Fact Sheet 196–99, 2 p. http://pubs.er.usgs.gov/usgspubs/fs/fs19699.

Jurgens, B.C., Böhlke, J.K., and Eberts, S.M., 2012, Tracer LPM (Version 1): An Excel® workbook for interpreting groundwater age distributions from environmental tracer data: U.S. Geological Survey Techniques and Methods Report book 4, chap. F3, 60 p.

Kauffman, L.J., and Chapelle, F.H., 2010, Relative vulnerability of public supply wells to VOC contamination in hydrologically distinct regional aquifers: Groundwater Monitoring and Remediation, v. 30, no. 4, p. 54–63, http://dx.doi.org/10.1111/j.1745–6592.2010.01308.x.

Kauffman, L.J., and Crandall, C.A., 2008, Use of observations of tritium and sulfur hexafluoride (SF6) concentrations to improve calibration of a flow and advective transport model, in Poeter, E.P., Zheng, Chunmio, and Hill, M.C., eds., Proceedings of MODFLOW and More 2008: Ground Water and Public Policy Conference, Golden, Colo., May 19–21, 2008, p. 395.

Keating, E.H., Vesselinov, V.V., Kwicklis, Edward, and Lu, Zhiming, 2003, Coupling basin- and site-scale inverse models of the Española Aquifer: Ground Water v. 41, no.2, p. 200–211.

Konikow, L.F., 2011, The secret to successful solute-transport modeling: Ground Water, v. 49, no. 2, p.144–159, http://dx.doi.org/10.1111/j.1745-6584.2010.00764.x.

Konikow, L.F., Hornberger, G.Z., Halford, K.J., and Hanson, R.T., 2009, Revised multi-node well (MNW2) package for MODFLOW groundwater flow model: U.S. Geological Survey Techniques and Methods, book 6, chap. A30, 67 p.

LaVenue, Marsh, Andrews, R.W., and Ramarao, B.S., 1989, Groundwater travel time uncertainty analysis using sensitivity derivatives: Water Resources Research, v. 25, no.7, p. 1551–1566.

Leake, S.A., Pool, D.R., and Leenhouts, J.M., 2008, Simulated effects of groundwater withdrawals and artificial recharge on discharge to streams, springs, and riparian vegetation in the Sierra Vista Subwatershed of the Upper San Pedro Basin, southeastern Arizona: U.S. Geological Survey Scientific Investigations Report 2008–5207, 14 p.

Leake, S.A. Reeves, H.W., and Dickinson, J.E., 2010, A new capture fraction method to map how pumpage affects surface water flow: Ground Water, v. 48, no. 5, p.690–700.

LeBlanc, D.R., 1984a, Digital model of solute transport in a plume of sewage contaminated ground water, in LeBlanc, D.R., ed., 1984, Movement and fate of solutes in a plume of sewage-contaminated water, Cape Cod, Massachusetts: U.S. Geological Survey Open-File Report 84–475, p. 11–44.

LeBlanc, D.R.,1984b, Sewage plume in a sand and gravel aquifer, Cape Cod, Massachusetts: U.S. Geological Survey Water-Supply Paper 2218, 28 p.

Lindgren, R.J., Houston, N.A., Musgrove, Marylynn, Fahlquist, L.S., and Kauffman, L.J., 2011, Simulations of groundwater flow and particle-tracking analysis in the zone of contribution to a public-supply well in San Antonio, Texas: U.S. Geological Survey Scientific Investigations Report 2011–5149, 93 p.

Mapa, R., Illangasekare, T.H., and Garcia, J.E., 1994, Upscaling of water flow and solute transport in saturated porous media: theory, computation, and experiments—Progress report submitted to the US Army Waterways Experiment Station: Mississippi, US Army.

Markstrom, S.L., Niswonger, R.G., Regan, R.S., Prudic, D.E., and Barlow, P.M., 2008, GSFLOW-Coupled ground-water and surface-water FLOW model based on the integration of the Precipitation-Runoff Modeling System (PRMS) and the Modular Ground-Water Flow Model (MODFLOW-2005): U.S. Geological Survey Techniques and Methods book 6, chap. D1, 240 p.

McMahon, P.B., Burow, K.R., Kauffman, L.J.,Eberts, S.M., Böhlke, J.K. and Gurdak, J.J., 2008, Simulated response of water quality in public supply wells to land use change: Water Resources Research, v. 44, no. 7, 16 p., http://www.agu.org/pubs/crossref/2008/2007WR006731.shtml.

Mehl, Steffen, and Hill, M.C., 2000, A comparison of solute-transport solution techniques based on inverse modeling results, in Stauffer, F., Kinzelbach, W., Kovar, K., and Hoehn, E., eds., Calibration and reliability in groundwater modeling: Coping with uncertainty: Wallingford, Oxfordshire, United Kingdom, International Association of Hydrological Sciences Press, p. 205–212.

Mehl, Steffen, and Hill, M.C., 2001, A comparison of solute-transport solution techniques and their effect on sensitivity analysis and inverse modeling results: Ground Water, v. 39, no. 2, p. 300–307.

Mehl, Steffen, and Hill, M.C., 2002, Development and evaluation of a local grid refinement method for block-centered finite-difference groundwater models using shared nodes: Advances in Water Resources, v. 25, p. 497–511.

Mehl, Steffen, and Hill, M.C., 2003, Locally refined block-centered finite-difference groundwater models: evaluation of parameter sensitivity and the consequences for inverse modeling, in Kovar, K., and Hrkal, Z., eds, Calibration and reliability in groundwater modeling: A few steps closer to reality: International Association of Hydrological Sciences Press, p. 227–232.

Mehl, Steffen, and Hill, M.C., 2005, MODFLOW-2005, The U.S. Geological Survey Modular Groundwater Model—Documentation of shared node local grid refinement (LGR) and the boundary flow and head (BFH) package: U.S. Geological Survey Techniques and Methods book 6, chap. A12, 68 p.

Mehl, Steffen, and Hill, M.C., 2007, MODFLOW-2005, The U.S. Geological Survey modular groundwater model—Documentation of the multiple-refined-areas capability of local grid refinement (LGR) and the boundary flow and head (BFH) package: U.S. Geological Survey Techniques and Methods book 6, chap. A21, 13 p., http://water.usgs.gov/nrp/gwsoftware/modflow2005_lgr/mflgr.html.

Mehl, Steffen, Faunt, C.C., Lacznaik, R., Li, Z., and Hill, M.C., 2006. Examination of groundwater pumping effects using regional-scale, site-scale, and locally refined numerical models, MODFLOW and MORE 2006, in Poeter, E.P., Hill, M.C., and Zheng, Chunmio, eds., Proceedings of MODFLOW and More 2006 Conference: Managing Ground-Water Systems Conference, Golden, Colo., May 21–24, 2006 (ver. 2), p. 529–533.

Mehl, Steffen, 2008, Coupling MODFLOW-LGR with SFR to represent stream-aquifer interactions, in Poeter, E.P., Zheng, Chunmio, and Hill M.C., eds., Proceedings of MODFLOW and More 2008 Conference: Ground Water and Public Policy Conference, Golden, Colo., May 19–21, 2008, p. 506–509.

Muir, K.S., and Coplen, T.B., 1981, Tracing groundwater movement by using the stable isotopes of oxygen and hydrogen, upper Penitencia Creek alluvial fan, Santa Clara Valley, California: U.S. Geological Survey Water-Supply Paper 2075, 18 p.

Newhouse, M.W., and Hanson, R.T., 2000, Application of three-dimensional borehole flow measurements to the analysis of seawater intrusion and barrier injection systems, Los Angeles, California: Proceedings of the Minerals and Geotechnical Logging Society annual meeting, October 2000, Golden, Colo., p. 281–292.

Newhouse, M.W., and Hanson, R.T., 2002, Three-dimensional flow measurements of ground water in uncased wells completed in volcanic basalts, Mountain Home Air Force Base, Idaho: U.S. Geological Survey Water-Resources Investigation Report 01–4259, 13 p. http://id.water.usgs.gov/PDF/wri014259/mhafb.pdf.

Newhouse, M.W., Hanson, R.T., Wentworth, C.M., Everett, R., Williams, C. F., Tinsley J., Noce, T. E., and Carkin, B.A., 2004, Geologic, water-chemistry, and hydrologic data from multiple-well monitoring sites and selected water-supply wells in the Santa Clara Valley, California, 1999–2003: U.S. Geological Survey Scientific Investigations Report 2004–5250, 134 p. http://pubs.water.usgs.gov/sir2004-5250/.

Niswonger, R.G., and Prudic, D.E., 2005, Documentation of the Streamflow-Routing (SFR2) Package to include unsaturated flow beneath streams—a modification to SFR1: U.S. Geological Techniques and Methods book 6, chap. A13, 50 p.

Niswonger, R.G., Prudic, D.E., and Regan, R.S., 2006, Documentation of the Unsaturated-Zone Flow (UZF1) Package for modeling unsaturated flow between the land surface and the water table with MODFLOW-2005 (rev. 2010): U.S. Geological Survey Techniques and Methods, book 6, chap. A19, 62 p.

Phillips, F.M., Tansey, M.K., Peeters, L.A., Cheng, S., and Long, A., 1989, An isotopic investigation of groundwater in a central San Juan basin, New Mexico: Carbon-14 dating as a basis for numerical flow modeling: Water Resources Research, v. 25, p. 2259–2273.

Plummer, L.N., Bexfield, L.M., Anderholm, S.K., Sanford, W.E., and Busenberg, Eurybiades, 2004, Hydrochemical tracers in the middle Rio Grande Basin, USA: 1. Conceptualization of groundwater flow: Hydrogeology Journal, v. 12, no. 4, p. 359–388.

Poeter, E.P., and Gaylord, D.R., 1990, Influence of aquifer heterogeneity on contaminant transport at the Hanford Site: Ground Water, v., 28, no. 6, p. 900–909.

Poeter, E.P., and Hill, M.C., 2008, SIM_ADJUST—A computer code that adjusts simulated equivalents for observations and predictions: International Ground Water Modeling Center report BWMI 2008-1, 28 p.

Poeter, E.P., Hill, M.C., Banta, E.R., Mehl, Steffen, and Christensen, Steen, 2005, UCODE-2005 and six other computer codes for universal sensitivity analysis, calibration, and uncertainty evaluation: U.S. Geological Survey Techniques and Methods book 6, chap. A11, 283 p.

Pollock, D.W., 1989, Documentation of computer programs to compute and display pathlines using results from the U.S. Geological Survey modular three-dimensional finite-difference groundwater flow model: U.S. Geological Survey Open-File Report 89–381, 188 p.

Pollock, D.W., 1994, User's guide for MODPATH/ MODPATH-PLOT, version 3: a particle tracking post-processing package for MODFLOW, the U.S. Geological Survey finite-difference groundwater flow model: U.S. Geological Survey Open-File Report 94–464, 249 p.

Reichard, E.G., Land, Michael, Crawford, S.M., Johnson, Tyler, Everett, R.R., Kulshan, T.V., Ponti, D.J., Halford, K.J., Johnson, T.A., Paybins, K.S., and Nishikawa, Tracy, 2004, Geohydrology, geochemistry, and groundwater simulation-optimization of the Central and West Coast Basins, Los Angeles County, California: U.S. Geological Survey Water Resources Investigations Report 2003–4065, 184 p.

Saltelli, Andrea; Ratto, Marco; Andres, Terry; Campolongo, Francesca; Cariboni, Jessica; Gatelli, Debora; Saisana, Michaela; and Tarantola, Stefano, 2008, Global sensitivity analysis, the primer: West Sussex, England, Wiley Interscience, 304 p.

Sanford, W.E., Plummer, L.N., McAda, D.P., Bexfielfd, L.M., and Anderholm, S.K., 2003, Use of environmental tracers to estimate parameters for a predevelopment groundwater flow model of the Middle Rio Grande basin, New Mexico: U.S. Geological Survey Water Resources Investigations Report 03–4286, 102 p.

Sanford, W.E., McAda, D.P., and Anderholm, S.K., 2004, Hydrochemical tracers in the middle Rio Grande Basin, USA: 2. Calibration of a groundwater-flow model: Hydrogeology Journal, v. 12, no. 4, p. 389–407.

Sanford, W.E., 2011, Calibration of models using groundwater age: Hydrogeology Journal, v. 19, no. 1, p. 13–16.

Schmid, Wolfgang; Hanson, R.T.; Maddock, Thomas, III; and Leake, S.A., 2006, User guide for the farm process (FMP1) for the U.S. Geological Survey's modular three-dimensional finite-difference ground-water flow model, MODFLOW-2000: U.S. Geological Survey Techniques and Methods book 6, chap. A17, 127 p., http://pubs.usgs.gov/ tm/tm6a17/.

Schmid, Wolfgang, and Hanson R.T., 2009, The Farm Process Version 2 (FMP2) for MODFLOW-2005—Modifications and Upgrades to FMP1: U.S. Geological Survey Techniques in Water Resources Investigations, book 6, chap. A32, 102 p., http://pubs.usgs.gov/tm/tm6a32/.

Shoemaker, W.B., Kuniansky, E.L., Birk, Steffen, Bauer, Sebastian, and Swain, E.D., 2007, Documentation of a conduit flow process (CFP) for MODFLOW-2005: U.S. Geological Survey Techniques and Methods, book 6, chap. A24, 50 p., http://pubs.usgs.gov/tm/tm6a24/.

Starn, J.J., Stone, J.R., and Mullaney, J.R., 2000, Delineation and analysis of uncertainty of contributing areas to wells at the Southbury Training School, Southbury, Connecticut: U.S. Geological Survey Water-Resources Investigations Report 00–4158, 54 p.

Starn, J.J., and Stone, J.R., 2004, Simulation of groundwater flow to assess geohydrologic factors and their effect on source-water areas for bedrock wells in Connecticut: U.S. Geological Survey Scientific-Investigations Report 2004–5132, 86 p.

Sykes, J.F., and Thomson, N.R., 1988, Parameter identification and uncertainty analysis for variably saturated flow: Advances in Water Resources, v. 11, p. 185–191.

Székely, Ferenc, 1998, Windowed spatial zooming in finite-difference ground water flow models: Ground Water, v. 36, no. 5, p. 718–721.

Thoms, R.B., Johnson, R.L., and Healy, R.W., 2006, User's guide to the variably saturated flow (VSF) process to MODFLOW: U.S. Geological Survey Techniques and Methods, book 6, chap. A18, 58 p., http://pubs.usgs.gov/tm/ tm6a18/.

Tiedeman, C.R., Hill, M.C., D'Agnese, F.A., and Faunt, C.C., 2003, Methods for using groundwater model predictions to guide hydrogeologic data collection, with application to the Death Valley regional ground-water flow system: Water Resources Research, v. 39, no. 1010, p. 5-1 to 5-17, http:// dx.doi.org/10.1029/2001WR001255.

Tiedeman, C.R., Ely, D.M., Hill, M.C., and O'Brien, G.M., 2004, A method for evaluating the importance of system state observations to model predictions, with application to the Death Valley regional groundwater flow system: Water Resources Research., v. 40, no. 12, 14 p., http://dx.doi. org/10.1029/2004WR003313.

von Rosenberg, D.U., 1982, Local mesh refinement for finite difference methods, Paper SPE 10974: Society of Petroleum Engineers Annual Fall Technical Conference, 57th, New Orleans, La., September, 26–29, 1982, http://dx.doi.org/.

Werner, A.D., Gallagher, M.R., and Weeks, S.W., 2006, Regional-scale, fully coupled modeling of stream-aquifer interaction in a tropical catchment: Journal of Hydrology, v. 238, no. 4, p. 497–510.

Wexler, E.J., 1992, Analytical solutions for one-, two-, and three-dimensional solute transport in groundwater systems with uniform flow: U.S. Geological Survey Techniques in Water Resources Investigations, book 3, chap. B7, 190 p.

Worthington, S.R.H., 2007, Groundwater residence times in unconfined carbonate aquifers: Journal of Cave and Karst Studies, v. 69, no. 1, p. 94–102.

Zheng, Chunmio, and Bennett, G.D., 1995, Applied contaminant transport modeling: theory and practice: New York, Van Nostrand Reinhold, 440 p.

Zhu, Chen, 2000, Estimate of recharge from radiocarbon dating of groundwater and numerical flow and transport modeling: Water Resources Research, v. 36, no. 9, p. 2607–2620.

Zimmerman, D.A., Hanson, R.T., and Davis, P.A., 1991, A comparison of parameter estimation and sensitivity analysis techniques and their impact on the uncertainty in ground water flow model predictions: U.S. Nuclear Regulatory Commission NUREG/CR-5522, Sandia National Laboratories SAND90-0128, Albuquerque, N.M., variously paged.

Appendix A. Input Instructions

MODPATH-OBS input is organized into blocks. Many users have become familiar with input blocks constructed as described here through their use in UCODE-2005 and other programs. For new users, the detailed description of input block found at the end of this appendix may be helpful. For example, table A2 notes that there can be no blank lines within input blocks and that comments can be included using "#".

The input blocks are demonstrated by using examples. The examples in the following input files are taken from the hypothetical example problem as much as possible.

Running MODFLOW, MODPATH, and MODPATH-OBS Using a Batch File

The batch files used to run MODFLOW-LGR and MODPATH-LGR for the example problem is called MF_RUN.bat. It runs the sequence of programs needed to produce a forward run of the model, including all particle tracking (fig. 11). This type of batch file is typically run by clip-on model analysis programs like PEST and UCODE_2005, and contains the sequence of programs needed for each forward model run. This batch file can also be used for "stand-alone" forward runs. The sequence of programs in the batch file is:

```
mflgr modflow.lgr

MPLGR MP.in

MODPATH_OBS MPOBS.in
```

The MODFLOW-LGR (MFLGR), MODPATH-LGR (MPLGR), and MODPATH-OBS (MODPATH_OBS) are run in sequence for each forward run, with each program specifying an input file for control and file information.

Modflow.lgr is the control file for MODFLOW-LGR, and for the example problem contains the following lines:

```
LGR                    ;Indicates this is an LGR input file
3                      ;# of grids
parent\parent.nam          ;Name file of parent model
PARENTONLY             ;GRIDSTATUS: (List parent first)
70 71                  ;Unit #'s for saving BFH (Boundary-Flow Head) info
child1\child1.nam          ;Name file of child model 1
CHILDONLY              ;GRIDSTATUS
1 -59 80 81            ;starting heads, IBOUND flag, unit #'s for BFH info
50  -1                 ;MXLGRITER, IOUTLGR: max. # of LGR iterations, print flag
0.40000 0.40000        ;RELAXH, RELAXF: relaxation for heads and fluxes
1.0E-5  1.0E-5         ;HCLOSELGR, FCLOSELGR: closure criteria for head and fluxes
1 20 22                ;NPLBEG,NPRBEG,NPCBEG: beginning layer, row and column
3 31 39                ;NPLEND,NPREND,NPCEND: ending layer, row and column
9                      ;NCPP: # of child cells per width of parent
3 3 3                  ;NCPPL (NPLBEG to NPLEND): # of child cells per parent layer
child2\child2.nam          ;Name file of child model 2
CHILDONLY              ;GRIDSTATUS
1 -60 92 93            ;starting heads, IBOUND flag, unit #'s for BFH info
50  -1                 ;MXLGRITER, IOUTLGR: max. # of LGR iterations, print flag
0.40000 0.40000        ;RELAXH, RELAXF: relaxation for heads and fluxes
1.0E-5  1.0E-5         ;HCLOSELGR, FCLOSELGR: closure criteria for head and fluxes
1 10 46                ;NPLBEG,NPRBEG,NPCBEG: beginning layer, row and column
3 21 63                ;NPLEND,NPREND,NPCEND: ending layer, row and column
9                      ;NCPP: # of child cells per width of parent
3 3 3                  ;NCPPL (NPLBEG to NPLEND): # of child cells per parent layer
```

MP.in is the MODPATH-LGR control file and for the example problem contains the following lines:

```
modflow.lgr
parent\parent.rsp          ;Parent response File
parent\flowp               ;Parent global output file
child1\child1.rsp          ;Child response file
59
child2\child2.rsp          ;Child response file
60
ENDPOINT 65 global_endpoint.dat
PATHLINE 67 global_pathline.dat
```

MPOBS.in, the MODPATH-OBS control file, is described in this Appendix and an example is shown below.

Parameter-Estimation Batch File

The parameter estimation batch input file for this example, ucode_run_exmpl.bat, runs UCODE-2005 with a UCODE main input file called exmpl_trck.in. A similar batch file could be constructed for PEST. The file ucode_run_exmpl.bat contains for the example problem considered here contains the following lines.

```
bin\ucode_2005.exe ucodeinput\exmpl_trck_ucode.in ucodeoutput\sens_all
rem > ucode_run_mpobs.log
pause
```

The identified input file, here exmpl_trck.in, contains all of the input required to perform forward runs, sensitivity runs, and parameter estimation runs. A file used in the example problem is shown in appendix B.

MODPATH-LGR Output File Read by MODPATH-OBS

When the line MPLGR MP.in is executed, a file intended to be read by MODPATH-OBS is produced. This file is largely the same as the output files from previous versions of MODPATH, but some columns have been added.

Items previously included in the MODPATH output file

The MODPAH-LGR output file read by MODPATH-OBS contains 23 columns of data.
1. Zone code for the cell containing the final location of the particle
2. J (column) index for the cell containing the final location
3. I (row) index for the cell containing the final location
4. K (layer) index for the cell containing the final location
5. Global coordinate in the x-direction (J index direction) for the final location
6. Global coordinate in the y-direction (I index direction) for the final location
7. Global coordinate in the z-direction (K index direction) for the final location
8. Local coordinates for the z-direction within the grid cell (0 to 1 within the model layer, -1 to 0 within an underlying confining layer)
9. Total tracking time
10. Global coordinate in the x-direction for starting location
11. Global coordinate in the y-direction for starting location
12. Local coordinates in the z-direction within the cell for starting location
13. J index for cell containing starting location
14. I index for cell containing starting location
15. K index for cell containing starting location
16. Zone code for cell containing starting location
17. Cumulative MODFLOW time step number corresponding to the time of release
18. Particle termination code, IPCODE
19. Release time

Items added at the end of each line for MODFLOW-LGR

20. Regional coordinate in the z-direction (K index direction) for the starting location
21. Identification number IDMODEL_START of the model where the particle started. For particles that start in the parent model, IDMODEL_START equals one. IDMODEL_START equals two if the particle starts in the first child model listed in the MODPATH-LGR control file. IDMODEL_START increases by one for each additional child model listed in the MODPATH-LGR control file.
22. Identification number IDMODEL of the model having the final particle endpoint. The parent model is number 1 and each subsequent child model specified in the MODPATH-LGR control file is numbered in ascending order beginning with number 2.
23. Particle identification number IDPART assigned in the starting locations file
24. Particle attributes PARTAS specified by the user in the starting locations file
    Once this and other MODPATH-LGR output files have been created, MODPATH-OBS can be run.

MODPATH-OBS Input Instructions

MODPATH-OBS Control File

The instructions needed to run the MODPATH-OBS program are supplied by the MODPATH-OBS control file. The name of this file can be supplied as an argument in the command line prompt, as in the batch file described in the beginning of this appendix. The main output file can be specified as an argument as well. The line is repeated here.

```
MPATH_OBS.exe InputFilename OutputFilename
```

If the InputFilename is not supplied as an argument the program will prompt for the filename. The default input filename, MODPATH_OBS_INPUT_FILE.txt, is specified by simply hitting ENTER at the prompt. If the output filename is not supplied, the default output filename, Modpath_Obs.out, will be used. The MODPATH-OBS control file is structured using Jupiter-API input blocks (Banta and others, 2006). Some input block basics are included after the description of the input blocks. Those familiar with the terms input block, blocklabel, and keyword will probably be able to understand the following input instructions without reviewing this material.

Table A1 lists the data input blocks that compose the MODPATH-OBS control file. A description of the contents of the each block is presented below with the corresponding example input from the related example model, if available, and a generic example, if not.

MODPATH-OBS Input blocks

Options Input Block (Required)

The Options input block can be used to control what is written to the main output. The keywords are usually read using the Keywords block format.

Verbose: Flag that controls what is written to the MODPATH-OBS main output file as follows. The default value of Verbose is 3.

Verbose	Output
0	No extraneous output
1	Warnings
2	Warnings, notes
3	Warnings, notes, echo selected input
4	Warnings, notes, echo all input. Includes all values read from process-model output files.
5	Warnings, notes, echo all input, plus some miscellaneous information. Includes all values read from process-model output files

Table A1. Blocklabels of the main input file for MODPATH-OBS. The shaded input blocks are used to define the geometry of the sources and observation location locations.

Blocklabel	Status	Purpose
Options	Required	Define application operation
Models	Required	List the Modpath name files for all models used in the simulation.
[1]Source_Zones_###	Optional. Repeat for models as needed	Define areas and volumes used to describe source locations in the Source_Poly input block.
[1]Observation_Zones_###	Optional. Repeat for models as needed	Define areas and volumes used to describe observation locations in Observation_Poly input block.
Source_Point	Optional	Defines source locations that are points or lines, possibly offset from the cell center.
Source_Poly	Optional	Associate source location identifiers to areas and volumes defined in the Source_Zones_### input blocks.
Observation_Point	Optional	Define observations locations that are points or lines, possibly offset from the cell center.
Observation_Poly	Optional	Associate source location identifiers to areas and volumes defined in the Observation_Zones_### input blocks.
Observation_Groups	Optional	Define values for observation data by group
Observation_Data	Required	Define name, type, source and observation locations, and other options for each observation.
Concentration_Files	Optional	List the concentration files
Exceedance_Definition	Optional	Define options for concentration observations with the exceedance option
Output_Files	Required	Define output and associated file names.

[1] Replace ### by parent, child1, child2, and so on, to identify the model to which the information pertains. If a single grid is used, replace ### by 1.

Time_Units: The time units used in specifying the sample time for the observations. Units will be converted if these units do not to match the units used in the MODPATH endpoint file as defined in the discretization file. Consistency is achieved by allowing the user time specification for the input and output from MODPATH-OBS. If this keyword is not included, no conversion will be done and the input should be in the same units as defined in the MODPATH discretization file. Options are:
–seconds
–minutes
–hours
–days
–years

Reference_Time: Time in the units of Time_Units for which the Modpath time equals zero. The Modpath reference time is defined in the response file as a value of simulation time in decimal years with respect to the end of the simulation. In MODPATH-OBS time is intended to represent real time, even when particles are tracked backwards in time. So, if the reference time is the year 2010, backward tracking of particles would produce earlier times such as 2005 and 2000. This is different than a backward tracking simulation in MODPATH where backward tracking is used and increasing values of time are back in time. The default value of Reference_Time is 0.

Modpath_Control: Name of the file that controls the performance of MODPATH or MODPATH-LGR. When there are multiple models that form a locally refined model, this is the MODPATH-LGR control file, for which an example file named mp.in is shown above in this appendix. When there is a single model and no local grid refinement, this is a MODPATH response file. This keyword is required, if it is not present, the program will be stopped.

TimeThresh: In the endpoint file used by this version of MODPATH-OBS, particle locations are defined when they reach a boundary or the end of the simulation time. In comparing the time specified in the endpoint file to times specified when defining observations, TimeThresh is used to determine how close is close enough. The comparison involved is between SampleTime (from the Observation_Data block) and the following:

- For forward tracking:
  The time defined by keyword Reference_Time in this input block plus the final time from the endpoint file (converted to the time units defined by the Time_Units keyword of this input block)

- For backward tracking:
  The time defined by keyword Reference_Time in this input block minus the release time from the endpoint file (converted to the time units defined by the Time_Units keyword of this input block)

TimeThresh needs to be specified using the units defined by the Time_Units keyword of this input block. The value should be chosen based on the precision of the times in the MODPATH endpoint file and the time between particle releases. Choosing time units in MODFLOW and MODPATH to avoid very large or small fractional values of time will help minimize problems with matching times. For example, do not use seconds as the time units for a 100 year simulation or years for a 1 day simulation. The default value of TimeThresh is 0.1.

Endpoint_Version: Version of MODPATH endpoint file. The default value of Endpoint_Version is Standard. Options are:
- Standard – The standard endpoint file from MODPATH-LGR or MODPATH version 5 and earlier.
- Compact – The compact endpoint file from MODPATH-LGR or MODPATH version 5 and earlier.
- Binary – The binary endpoint file from MODPATH-LGR or MODPATH version 5 and earlier.
- MODPATH6 – The endpoint file from MODPATH version 6

Volume_Column: Column index in the endpoint file that contains the volume of water that each particle represents. This volume, if specified, will be used in weighting any averaging that is done for particles. MODPATH-LGR and MODPATH 6 allow for a label in the starting point file that will be appended at the end of each line in the endpoint file, so a volume could be put there. Alternatively, a separate program could be run between running MODPATH and MODPATH-OBS that could add this column. A value of zero indicates that volume will not be read or used in calculations and the assumed volume for each particle is equal. The default value of Volume_Column is 0.

Example of an Options input block using keyword format:

```
Begin Options Keywords
  Verbose=5
  Run_Type=UCODE
  Time_Units=Years
###Define the reference time at the end of the simulation because particles
## are tracked backwards in time.
  Reference_Time=2010
  Modpath_Control_File=mp.in
  TimeThresh=0.1
End Options
```

Models Input Block (Required)

The Models input block lists the MODPATH-LGR name files used to run the parent model and then any child models used in the simulation. The order needs to match the order of the models listed in the MODPATH-LGR files, as defined in the MODFLOW-LGR and MODPATH-LGR input files listed at the beginning of this appendix.

The two keywords associated with the Models input block and are usually read arranged in Table format with NROW being the number of models and NCOL = 1 or 2 (2 if the Model_ID column is included; 1 if not).

Modpath_Namefile: The name file for each model as defined for the related MODPATH-LGR model run. This is required for each model.

Model_ID: This is an optional item used to reference models within this input file. The name assigned can be from 1 to 16 characters in length and cannot contain spaces. If not defined, Model_ID is set to Model#, where # is replaced by a number that equals the sequential order that the model is listed in the Models input block. Thus, if there is no local grid refinement, one grid is used, and Model_ID is not defined, Model_ID is set to 1.f

Example of a Models input block using keyword format:

```
Begin Models Table
  Nrow=3 ncol=2 ColumnLabels
  Modpath_Namefile          Model_ID
  MODPATH_PARENT.dat        Parent
  MODPATH_CHILD1.dat        Child1
  MODPATH_CHILD2.dat        Child2
END Models
```

For example, the Modpath_Parent. Dat file might contain the following lines:

```
main        10      parent\parent_main.dat
dis         15      parent\parent.dis
budget      17      parent\flowp2
head(BINARY) 18     parent\hedp
endpoint     21     parent\parent_endpoint.dat
pathline     23     parent\parent_pathline.dat
locations    22     parent\parent_start3.dat
```

Defining Source and Observation Locations

There are options to define source and observation locations using volumes, areas, lines, or points, as shown in figure 2A. Source locations are always defined as oldest in time, while observation locations are always defined as newest in time. Particles are associated with source and observation locations based on location; they are not associated based on particle numbers or names defined in MODPATH.

These characteristics produce the following performance of MODPATH-OBS:

- Forward particle tracking tracks particles from source locations to observation locations. Particles are released at the source location.

- Backward particle tracking tracks particles from observation locations to source locations. Particles are released from the observation location.

- In both cases, the particles involved are obtained by MODPATH-OBS by comparing the source or observation location definition to the particle release locations defined in the MODPATH endpoint output file.

Different characteristics may be applied to the same particles. For example, when used in a proximity observation, only the final particle locations as read from the MODPATH endpoint output file are of consequence. When used in a concentration observation, the particles also can be assigned starting concentrations which can be changed based on the travel time and decay rate and only the particles that reach a given destination may be of interest.

Source and observation locations are defined using the input blocks listed in table A1. The input blocks need to be ordered as shown in table A1, and are described in that order. The general utility of the input blocks is described in the following paragraphs.

The Source_Zones_### and Observation_Zones_### input blocks are used to define zones in the one or more models used to simulate the system. Replace the ### with the corresponding Model_ID from the Models Input Block. If that keyword is omitted in the Model input block, use sequential number, where 1 is the first model listed in the Models input block, 2 is the second, and so on. The zone number can be used in the Source_Poly and Observation_Poly input blocks listed later in the input file.

The Source_Point and Observation_Point input blocks provide convenient ways to define point and line source and observation locations. For example, a vertical line might be used to represent a well screen. These input blocks provide the opportunity to define a point anywhere within a model cell, and a vertical line at any areal location and of any length up to the thickness of the system. These input blocks are best used when the sources or sinks are relatively weak. Generally particles should be started from locations represented by points or lines (forward tracking from source locations, backtracking from observation locations) because particles that terminate normally in Modpath are stopped at cell boundaries. With the exception of proximity observations, observations based on transport from a point or line source location to a point or line observation location will not work.

The Source_Poly and Observation_Poly input blocks are used to define areas and volumes. The areas and volumes are always bounded by grid cell faces. Generally these volumes are a good choice to represent pumping/injection wells and other internal sources or sinks. Areas are a good choice to represent recharge, evapotranspiration, surface water features, and any other flows that can be assigned to a cell face on the boundary of the model. These input blocks use zones defined in the Source_Zones_### and Observation_Zones_### input blocks. The Source_Poly and Observation_Poly input blocks allow the user to define the bounding cell face.

Source_Zones_### Input Block (Optional; Repeat for models as needed)

The Source_Zones_### input block can define zones using lists of cells, or using files that contain arrays similar to MODFLOW IBOUND arrays. Both methods can be used to define zones in a single input block. The last zone that is assigned to a cell will be used.

In the input block name, ### is replaced by a model name as defined in the Models input block, or by a sequential number if Model_ID is not defined. The layer, row, and column numbers listed in a Source_Zones_### input block are relative to that model.

The keywords associated with the Source_Zones_### input block are as follows.

Zone_ID: The number or name of the zone. Within the Source_Zones_### input blocks, repeated values of Zone_ID (or repeating zones defined in an array) mean that the defined zones are combined.

  If Array_Data is used, any number defined using the Zone_ID keyword will be replaced by numbers listed in the array. Alternatively, set ZoneID=array, and definition of Array_Data will be expected. If ZoneID=array and Array_Data is not defined, an error message is printed and execution of MODPATH-OBS stops.the By default value, Zone_ID=array.

KTOP: The model layer for which the layer top defines the top of the zone. This keyword is required if this block is used, if it is not present, the program will be stopped.

KBOT: The model layer for which the layer bottom defines the bottom of the zone. If the keyword KBOT is not included, the value is set to KTOP.

Array_Data: Literal string with array reading instructions defined using the Constant or Open/Close formats from MODFLOW (Harbaugh, 2005, p. 8-57 to 8-59), repeated here. The literal string needs to be enclosed in double quotes. Nonzero values listed in the array are used to define one or more occurrences of Zone_ID. That is, an array with non-zero value 2, 3 and 4 produces three values of Zone_ID that can be used in Source_Poly to define sources.

CONSTANT CNSTNT
All values in the array are set equal to CNSTNT. CNSTNT needs to be an integer; no decimal point can be used. Place the entire statement in double quotes.

OPEN/CLOSE FNAME CNSTNT FMTIN IPRN
FNAME: The name of the file from which the array is read. Numbers in the array are used as the Zone_id. Within a single zone, repeat the same integer for each cell. The file is read from the directory from which MODPATH-OBS is executed unless defined using a path. Paths may be relative or absolute. Place the entire statement in double quotes.

FMTIN: The Fortran format statement used to read the file. Commonly the array is space delimited, and FMTIN can be set to (FREE).

CNSTNT: All values read are multiplied by CNSTNT. Here, CNSTNT generally is set to 1.

IPRN: An integer that controls how the array is printed. If IPRN is a negative integer, the array is not printed. IPRN set to 2 results in up to 40 values printed on each line, with each number given two spaces. IPRN set to 3 results in 30 values printed on each line, with each number given three spaces. Lines are repeated until values for all grid columns are printed.

JCOL: the lower column number of the zone. If ArrayData is set to NONE, this keyword is required, if it is not present, the program will be stopped.

JCOL2: the upper column number of the zone. This keyword allows for a range of cells to be used in defining the zones defining the upper limit of the range. If the value of JCOL2 is less than JCOL, the values will be reversed. If this keyword is not defined it will be set to JCOL.

IROW: the row number of the zone. If ArrayData is set to NONE, this keyword is required, if it is not present, the program will be stopped.

IROW2: the upper row number of the zone. This keyword allows for a range of cells to be used in defining the zones defining the upper limit of the range. If the value of IROW2 is less than IROW, the values will be reversed. If this keyword is not defined it will be set to IROW.

An example of the Source_Zones_### input block is shown below with the Observation_Zones_### Input Block.

Observation_Zones_### Input Block (Optional. Repeat for models as needed)

Observation_Zones_### input blocks are constructed using the same instructions provided above for Source_Zones_### input blocks. The Zone_ID list is separate, so Zone_ID does not need to be coordinated with the designations used in the definition of Source_Zones_### in put blocks.

Example of the Source_Zones_#### and Observation_Zones_#### input blocks constructed using table format for multiple models, with ### replaced by Parent, Child1, or Child 2 as needed:

```
#########################################
# DEFINE SOURCE AND OBSERVATION ZONES #
#########################################

########### Parent Model ##############
Begin Source_Zones_Parent Table
  Nrow=1  ncol=3 ColumnLabels
  KTOP KBOT    Array_Data
  1    1     "Open/Close .\parent\parent_source_array_lay1.dat 1 (FREE) -9"
End Source_Zones_Parent

Begin Observation_Zones_Parent Table
  Nrow=4 ncol=5 ColumnLabels
  Zone_ID JCOL  IROW  KTOP KBOT
  2         3    3     1    1
  3         3   19     1    1
  4         3   34     1    1
  5         3   45     1    1
End Observation_Zones_Parent

###########Child Model 1##############
Begin Source_Zones_Child1 Table
  Nrow=1  ncol=3 ColumnLabels
  KTOP KBOT Array_Data
  1    1        "Open/Close .\child1\child1.srz 1 (FREE) -9"
End Source_Zones_Child1

Begin Observation_Zones_Child1 Table
  Nrow=1 ncol=5 ColumnLabels
  Zone_ID JCOL  IROW  KTOP  KBOT
  6         73   55    1     6
End Observation_Zones_Child1

###########Child Model 2##############
#NO SOURCES IN CHILD MODEL 2 SO SOURCE_ZONES_### BLOCK OMITTED

Begin Observation_Zones_Child2 Table
  Nrow=1 ncol=5 ColumnLabels
  Zone_ID JCOL  IROW  KTOP  KBOT
  7         73   55    1     6
End Observation_Zones_Child2
```

Source_Point Input Block (Optional)

The keywords for the Source_Point input block define locations where water (and perhaps solute) enters the groundwater system from a point or line. The Source_Point input block is usually read using the Table block format with NROW being the number of source locations being defined and NCOL = 10. The keywords for the Source_Point input block are defined as:

Loc_ID: The source location identifier. This may be a number or name and can include up to 80 characters. The characters can include a-z, A-Z, 0-9, @#$%^&*(). Spaces are not allowed. This keyword is required if this block is used, if it is not present, the program will be stopped.

MODEL_ID: Model_ID from the Models Input Block corresponding to the Source Point. This keyword is required if this block is used and there is than 1 model, if it is not present, the program will be stopped.

The following six keywords are required if this block is used and there, if any one of them is not present, the program will be stopped.

JCOL: Indicates the column number in the corresponding model.

IROW: States the row number in the corresponding model.

KTOP: The model layer for which the top defines the top of the observation.

LOCX: The offset along the row used to define the point location within the cell, JCOL, IROW. LOCX values range from 0.0 to 1.0. To specify the cell center, set LOCX=0.5.

LOCY: The offset along the column used to define the point location within the cell, JCOL, IROW. LOCY values range from 0.0 to 1.0. To specify the cell center, set LOCY=0.5.

LOCZT: The local vertical offsets used to define the top of the source location (for example, the top of a well screen) within the cell, JCOL, IROW, KTOP. LOCZT values range from 0.0 (cell bottom) to 1.0 (cell top). To specify the cell center set LOCZT=0.5. Default is 0.0.

The following three keywords are used to extend the point defined above to a line. The line must be aligned in the column, row, or layer direction of the model.

LineDir: The direction in which the line will extend. The default value of LineDir is Z. Valid options are:
- **X** (across multiple columns)
- **Y** (across multiple rows)
- **Z** (across multiple layers)

INDEX2: The upper model index limit of layer (LineDir=Z), column(LineDir=X), or row(LineDir=Y) for which a line location.

LOC2: The local offset in the direction defined by LineDir. LOC2 is used to define end of a line location (for example the bottom of a well screen) within the cell containing the end of the line. LOC2 values range from 0.0 (cell bottom, left, or front) to 1.0 (cell top, right, or back.). To specify the cell center in the LineDir direction set LOC2=0.5. Default is 0.0.

The following Source_Point input block is not used in the example problem, but provides an example of the input format.

```
Begin Source_Point Table
  Nrow=1  Ncol=10 ColumnLabels
  Loc_ID   MODEL_ID    JCOL   IROW KTOP LOCX LOCY LOCZT LINEDIR INDEX2 LOC2
  well     parent      54     40    1    0.1  0.4  0.8   Z        2     0.1
End Source_Point
```

Source_Poly Input Block (Optional)

The keywords for the Source_Point input block define locations where water (and perhaps solute) enters the groundwater system from an area or volume. The Source_Point input block is usually read using the Table block format with NROW being the number of observation locations being defined and NCOL = 3. The keywords for the Source_Point input block are defined as:

Loc_ID: Identifies the source location. This may be a number or a name and can include up to 80 characters. The characters can consist of a-z, A-Z, 0-9, or symbols such as -,.@#$%^&*(). This keyword is required if this block is used, if it is not present, the program will be stopped.

Zone_ID: Identifies a zone number from the arrays defined in the Source_Zone input block. This keyword is required if this block is used, if it is not present, the program will be stopped.

Face_No: The cell face number used to identify incoming or outgoing particles relative to the source observation. Values range from 0 to 6 and the convention is the same as that defined for MODPATH (Pollock, 1994, p. 3-9; see figure 2B of this report). *Note: For particles to stop at sources on external faces, use the IFACE auxiliary variable for external boundaries in* MODFLOW *together with the COMPACT BUDGET AUX option in the* MODFLOW *output control file. In addition, the IRCHTP variable in* MODPATH *should be consistent with the definition here (generally IRCHTP ≠ 0 and the face set to 6.)* The default value of Face_No is 0.

Number	Cell side	Number	Cell Side
1	Left face. Along row, toward lower column number	4	Back face. Along column, toward lower row number
2	Right face. Along row, towards larger column number	5	Bottom face. Toward larger layer number.
3	Front face. Along column, toward larger row number	6	Top face. Toward lower layer number
0	No face assigned. Source/sink is distributed in the cell. Particles approaching the cell are stopped at whatever cell boundary is approached.		

Example of a Source_Poly input block constructed using table format:

```
Begin Source_Poly Table
   Nrow=4  ncol=3 ColumnLabels
   Loc_ID   Zone_ID   Face_No
   pit          1         6
   trench       3         6
   river        4         6
   farm         5         6
End Source_Poly
```

Observation_Point Input Block (Optional)

The keywords for the Observation_Point input block define locations where water (and perhaps solute) leave the groundwater system from a point or line. The Observation_Point input block is usually read using the Table block format with NROW being the number of source locations being defined and NCOL = 10. Observation_Point input blocks are constructed using the same instructions provided above for Source_Points input blocks. The Loc_ID list is separate, so Loc_ID does not need to be coordinated with the designations used in the definition of Source_Point in put blocks although it should be coordinated with the Observation_Poly input block.

Example of an Observation_Point input block constructed using table format:

```
Begin Observation_Point Table
  Nrow=1 ncol=10 ColumnLabels
  Loc_ID    MODEL_ID  JCOL IROW KTOP LOCX LOCY LOCZT INDEX2 LOC2
  well1_obs  parent     54   40    1   0.1  0.4  0.8    2    0.1
End Observation_Point
```

Observation_Poly Input Block (Optional)

The keywords for the Observation_Point input block define locations where water (and perhaps solute) leaves the groundwater system from an area or volume. The Observation_Point input block is usually read using the Table block format with NROW being the number of observation locations being defined and NCOL = 3. Observation_Poly input blocks are constructed using the same instructions provided above for Source_Poly input blocks. The Loc_ID list is separate, so Loc_ID does not need to be coordinated with the designations used in the definition of Source_Point in put blocks although it should be coordinated with the Observation_Point input block. Example of an Observation_Poly input block:

```
Begin Observation_Poly Table
  Nrow=6 ncol=3 ColumnLabels
  Loc_ID      Zone_ID     FACE_NO
  well2_obs   6              0
  well3_obs   7              0
  rivrch1_ob  2              0
  rivrch2_ob  3              0
  rivrch3_ob  4              0
  rivrch4_ob  5              0
End Observation_Poly
```

Observation_Groups Input Block (Optional)

Use the Observation_Groups input block to assign data that apply to all or many of the observations within assigned groups. Data for individual observations can be assigned in the subsequently read Observation_Data input block. When quantities are specified in both input blocks, data specified in the Observation_Data input block are used. Keywords in the input block include:

GroupName: Name for a group of observations (up to 12 characters; not case sensitive). Default=DefaultObs.

Other keywords: Any keywords from the Observation_Data input block.

Observation_Data Input Block (Required)

Information defining the observations that will be generated by the MODPATH-OBS program is entered in the Observation_Data input block. The Observation _Data input block is usually read using the Table block format with NROW being the number of source locations being defined and NCOL = 10.

There is also a UCODE-2005 input block called Observation_Data. Because undefined keywords are ignored in the JUPITER API used to construct UCODE_2005, and the input blocks of MODPATH-OBS, the same Observation_Data input block can be used for both programs. This reduces the chance for input error and was used to create the Observation_Data input block constructed for the hypothetical test case.

The keywords defined by MODPATH-OBS for the Observation_Data input block follow. Some are the same as those listed for the UCODE-2005 Observation_Data input block.

ObsName: States the name of the observation. This keyword is required; if it is not present, the program will be stopped.

ObsType: Indicates the observation type. This keyword is required; if it is not present, the program will be stopped. Valid values are:

- Proximity–Observation that particles in the observation location originated at a specified source location

- Time–Observation of the travel time from the source location to the observation location

- Conc–Observation of a concentration of a specified solute at the observation location

- Source–Observation that a percentage of the water in the observation location originated from the specified source location

Comp_Type: This keyword has different valid values depending on the ObsType that indicates the computation that will be done to generate the simulated value. This keyword is required; if it is not present, the program will be stopped.

For the ***Proximity ObsType***, a value is calculated for each particle as measured from the Observation Location (for forward tracking) or the Source Location (for backward tracking) at the sampling time. A value of zero means the particle has reached its desired location; see figure 5. To measure the distance from the starting location, define the Observation location to be the same as the source location. By default a total Cartesian distance (D) is used, and is calculated as $D = \sqrt{X^2 + Y^2 + Z^2}$. The following statistics can be used:

Cartesian distance	For the particle identified by Min, Med, or Max as determined using the total Cartesian path length, the x, y, or z component	The x-, y- or z-component from any particle
• Min – Minimum • Med – Median • Avg – Average • Max – Maximum	• MinX, MinY, MinZ • MedX, MedY, MedZ • MaxX, MaxY, MaxZ	• XMin, YMin, ZMin • XMed, YMed, ZMed • XAvg, YAvg, ZAvg • XMax, YMax, ZMax

When using Xmin, Ymin, and Zmin, the minimum values may come from different particles. The same is true for XMed, YMed, and ZMed, or XMax, YMax, and ZMax.

For the _**Time ObsType**_, a value is calculated for each particle associated with both the specified source and observation locations. For forward tracking, a value is calculated for particles that start at the defined Source Location and reach the defined Observation Location. For backward tracking, a value is calculated for particles tracked backward from the observation location that reach the Source Location. If no particle reaches the intended destination, observations are assigned the default value defined using the keyword NoPartValue in the Observation_Data input block. Values reported for time observations have units of the model unless the keyword specifies another time unit. Using the particles with calculated values, the following statistics can be reported as simulated values:

- Min–The minimum value
- Med–The median value
- Avg–The average value
- Max–The maximum value
- PctXX#–Used to indicate the percent of particles above, below, or equal to a threshold. Two letters replace XX to define the relation. The valid entries are:
  - LT–less than
  - LE–less than or equal to
  - EQ–equal to
  - GE–greater than or equal to
  - GT–greater than

The # represents the threshold against which the times are compared.

The Time_Units indicates the time units of #. The valid entries are

- b or blank   time unit of the model
- s   seconds
- m   minutes
- h   hours
- d   days
- w   weeks
- y   years

Here are some examples:

PctGT100   percent of travel times greater than 100

PctLT5   percent less than 5

For the _**Conc ObsType**_, a concentration is calculated for each particle associated with the observation location at the sample time. Currently all particles are assumed to represent equal amounts of water, so the average value of all particles is used as the simulated concentration value. With this observation type, the following values are valid:

- Conc–the simulated value is equal to the average concentration of all associated particles
- ExcDefID–The ExcDefID should be an entry from the ExceedanceDefinition table defined below. This indicates that the simulated value will be a percent of samples expected to be above or below a threshold as specified in the ExceedanceDefinition table. In this case the value of sample time is ignored, and a concentration is calculated for a range of times defined by the exceedance definition.

For the **Source ObsType,** the simulated value is the percent of particles associated with the observation location at the sample time that originate from the specified source type. The value for Comp_Type does not matter although is cannot be blank. It is recommended to use Percent in the Comp_Type column.

ObsLocID: Indicates the name of the observation location (as defined in either the Observation_Point or Observation_Poly input blocks) that the observation is from. This keyword is required; if it is not present, the program will be stopped.

ObsValue: The observed value. For the proximity ObsType, the observed value is assumed by definition to be 0. When transform = yes, the base 10 logarithm of the observed value should be entered. This keyword is required; if it is not present, the program will be stopped.

SampleTime: The time at which particles at the observation location are selected for the observation calculation. For backward tracking, this will be the release time and will be compared to the release time in the endpoint file to ensure consistency. For forward tracking, this will the final time and will be compared to the final time in the endpoint file to ensure consistency. The time comparisons use the value defined by the TimeThresh keyword in the Options input block. If the difference in the times exceeds TimeThresh, an error message will be printed and execution will stop. For steady state flow simulations, the sample time is important only for the concentration observations. This keyword is required; if it is not present, the program will be stopped.

SrcLocID: For the Proximity, Time, and Source ObsType, this indicates the name of the source location (as defined in either the Source_Point or Source_Poly input blocks) that the observation relates to. For the Time obstype, ALL can be used to do the requested calculation on all particles associated with the well without regard to their source location. For the Conc ObsType, this indicates the constituent that concentration will be calculated. The value entered should correspond to the name in at least one of the concentration files defined in the ConcentrationFiles input block. This keyword is required; if it is not present, the program will be stopped.

Transform: Allows for the simulated values to be transformed using a base-10 logarithm. Values of "yes" are only valid for the Time and Conc obstypes. The default value of Transform is no.

NoPartValue: The value that will be assigned in the output file if no particles are associated with the observation for the specified sampling time. The default value of NoPartValue is1E20.

PrintPart: Controls printing of particle information when XYZPRT, TIMPRT, CONPRT, TYPPRT are defined in the Output Files input block. Yes: include in the printing. No: do not include in the printing. The default value of PrintPart is yes.

GroupName: Group name from the Observation_Groups input block. The group attributes defined in the Observation_Groups input block are assigned to the observation and are then changed to attributes from the Observation_Data input block if specified. The default value of GroupName is DefaultObs.

Example of an Observation_Data input block:

```
BEGIN OBSERVATION_DATA TABLE
   Nrow=9 ncol=8 ColumnLabels
   ObsName      OBSType      Comp_Type    Obs10C_ID    ObsValue    SampleTime    Source_ID    Transform
   xyzobs1      Proximity    Xmed         well1_obs    0.000       2010          farm         no
   timmedpit2   Time         Med          well2_obs    33.996      2010          pit          no
   timmedpit2t  Time         Med          well2_obs    33.996      2010          pit          yes
   cnc_cfc1     Conc         Conc         well1_obs    372.467     2010          cfc          no
   cnc_pce1     Conc         Conc         well1_obs    0.000       2010          pce          no
   w2pceExc1    Conc         Exc1         well2_obs    0.677       1             pce          no
   w2pceExc2    Conc         Exc2         well2_obs    0.500       2             pce          no
   typfarmtol   Source       Percent      well1_obs    1.000       2010          farm         no
   typtrench3   Source       Percent      well3_obs    0.008       2010          trench       no
END OBSERVATION_DATA TABLE
```

Defining Concentrations (Optional)

Defining concentrations requires a Concentration_Files input block in the MODPATH-OBS main input file and additional files listed in the Concentration_Files input block. The additional files are composed of input blocks in the following order:

- Concentration_Instructions (Required in each concentration file.)
- Concentration_SourceZones (Optional. If omitted, the concentrations apply to all source zones)
- Concentration_History (Required in each concentration file.)

These input blocks are all described below and are followed by an example that includes all of them.

Concentration_Files Input Block (Optional)

The Concentration_Files input block specifies the files needed to define the concentration histories for source areas for which concentration observations are to be calculated. The only keyword is:

FileName: Name (including a path if needed) of the concentration file. This keyword is required if this block is used, if it is not present, the program will be stopped.

Each concentration file consists of two required input blocks, and one optional:

Information Input Block (Required for any file listed in the Concentrations input block)

In the Information input block, keywords include:

NAME: A name to define the concentration. This is used in the Source_ID entry in the Observation_Data input block to define how concentrations are computed. This keyword is required if this block is used, if it is not present, the program will be stopped.

DECAYORDER: The order of the equation used to define the decay of the initial concentration. Valid options are 0 or 1. If the keyword is omitted, the initial concentration is not decayed.

DECAYRATE: The rate constant used to define decay of the initial concentration. The decay rate is k of equations 1 and 2. DECAYRATE is ignored if the DECAYORDER keyword is omitted.

BREAKDOWNNAME: Name used to define a breakdown product of the main constituent defined by keyword NAME. This is used in the Source_ID entry in the Observation_Data input block to define how concentrations are computed. If keyword BREAKDOWNNAME is omitted, no breakdown products can be referred to.

BREAKDOWNRATIO: The amount of breakdown product produced for each unit of parent product degraded. Keyword BREAKDOWNRATIO is ignored if BREAKDOWNNAME is not defined.

SourceZones Input Block (Optional)

The SourceZones input block contains one keyword.

SOURCEZONE: Defines the name of a source zone (as defined by Loc_ID in either the Source_Point or Source_Poly input blocks) where the concentration defined in the file will be applied. If the keyword is not included, the concentration will be applied to all source areas.

Concentration_History Input Block (Optional)

The concentration history is generally defined using Table format. The table defines two keywords. The keywords are:

Time: Time for which the concentration is defined. Fractional times are permitted. The times should be presented in ascending order.

Concentration: Concentration at the associated time. Linear interpolation is used to calculate the concentration at times between the entered values to obtain values at times required by MODPATH-OBS.

Example of a Concentration_Files input block.

```
Begin Concentration_Files Table
  Nrow=3 ncol=1 ColumnLabels
  FileName
  MPObsInput\cfc.dat
  MPObsInput\pcepit1.dat
  MPObsInput\pcetrench1.dat
End Concentration_Files
```

Example of a Concentration File.

```
Begin Information Keywords
Name = PCE
DECAYRATE = 0.0100000000000
DECAYORDER = 1
End Information

Begin Source_Zones Keywords
Sourcezone=PIT
End Source_Zones

Begin Concentration_History Table
  NRow=5 NCol=2 ColumnLabels
  Time        Concentration
  -1000000        0
     1969.9       0
     1970        100
     1990        100
     1990.1        0
End ConcentrationHistory
```

Exceedance_Definition Input Block (Optional)

The Exceedance_Definition input block is needed if any of the concentration observations use the exceedance Comp_Type. The keywords for this input block are:

ExcDefID–Unique ID for the definition. Used in the Comp_Type column of the Observation_Data table. This keyword is required if this block is used, if it is not present, the program will be stopped.

StartTime–The earliest time for which concentration values will be calculated. This keyword is required if this block is used, if it is not present, the program will be stopped.

EndTime–The latest time for which concentration values will be calculated. This keyword is required if this block is used, if it is not present, the program will be stopped.

Threshold–The threshold value for which the exceedance will be calculated. This keyword is required if this block is used, if it is not present, the program will be stopped.

NObs–The number of observations that will make up the calculation. For transient simulations, particles will need to be released in conjunction with each observation. For steady state simulations, the concentrations will be calculated at equal intervals ([EndTime-StartTime]/[Nobs-1]) between and including the start and end times. This keyword is required if this block is used, if it is not present, the program will be stopped.

Operator–Determines how the threshold will be used. This keyword is required if this block is used, if it is not present, the program will be stopped. Valid options are:

- LT – less than
- LE – less than or equal to
- EQ – equal to
- GE – greater than or equal to
- GT – greater than

Example of an Exceedance_Definition input block.

```
Begin Exceedance_Definition Table
  Nrow=2 ncol=6 ColumnLabels
  ExcDefID    StartTime    EndTime    Threshold    NObs    Operator
    Exc1      1970         2000       5            31      gt
    Exc2      2001         2010       5            10      gt
End Exceedance_Definiton
```

Output_Files Input Block (Required)

The Output_Files input block specifies the files where output from the various observation types will be written. (Note: Instruction file(s) can be created by MODPATH-OBS, however they needed to be created before UCODE is run; thus when MODPATH-OBS is used to create the instruction files, it should first be run independent of UCODE.)

FType: The type of the output file; the 12 options are listed below. At least one of the first four needs to be listed to obtain results from MODPATH-OBS. This keyword is required if this block is used, if it is not present, the program will be stopped.

- XYZDAT–the simulated and observed values and residuals for proximity observations.
- TIMDAT–the simulated and observed values and residuals for time-of-travel observations.
- CONDAT–the simulated and observed values and residuals for concentration observations.
- SRCDAT–the simulated and observed values and residuals for source observations.
- XYZINS–the instruction file for UCODE or PEST for the proximity data.
- TIMINS–the instruction file for UCODE or PEST for the time-of-travel data.
- CONINS–the instruction file for UCODE or PEST for the concentration data.
- SRCINS–the instruction file for UCODE or PEST for the source data.

The following four options produce results for each particle associated with each observation. They are often used to resolve problems. The items printed in the file defined for each keyword are listed. The output can be limited to selected observation by adding a column titled PrintPart (yes or no, default is yes) in the Observation_Data input block.

- XYZPRT - obs_id, part_id, component_distance or total_distance (the component distance is the component x,y,z asked for by the observation, or total is just repeated if an individual component is not requested).
- TIMPRT - obs_id, part_id, travel_time. Information is printed only for particles that have reached the location for which time is desired.
- CONPRT - obs_id, part_id, recharge_time, travel_time, source_name, volume, concentration.
- SRCPRT - obs_id, part_id, yes or no if the particle is associated with the requested source type.

FILENAME: The file name to which the information defined by FType is printed. This keyword is required if this block is used, if it is not present, the program will be stopped.

Example of an Output_Files input block:

```
Begin Output_Files Table
  Nrow=8 ncol=2 ColumnLabels
  FType              FileName
  XYZDAT             MPObsOutput\xyz.dat
  TIMDAT             MPObsOutput\tim.dat
  CONDAT             MPObsOutput\conc.dat
  SRCDAT             MPObsOutput\src.dat
  XYZINS             UCodeInput\XYZUCODE.ins
  TIMINS             UCodeInput\TIMUCODE.ins
  CONINS             UCodeInput\CONUCODE.ins
  SRCINS             UCodeInput\SRCUCODE.ins
End Output_Files
```

General Input Block Instructions

The input blocks used in this work are programmed using modules from the JUPITER API (Banta and others, 2006). This section provides information about how the input blocks work.

Input blocks

The main input file includes input blocks with the basic structure:

```
Begin blocklabel [blockformat]
  Blockbody: many lines OR, when blockformat is 'files', a list of one or more files
End blocklabel
```

The brackets around blockformat indicate that it is optional. Square brackets are used to identify optional input throughout this document. All input is case-insensitive and space-delimited.

Blocklabel and blockformat are defined in the following sections. The definition of blockbody depends on blocklabel; the possible content of blockbody for each of the blocklabels available in MODPATH-OBS is described in the following sections.

The input blocks described in this report are part of the JUPITER API or are designed using the conventions established as part of the API.

Blocklabel

The variable blocklabel identifies the purpose of the data block and the data it can contain. This chapter provides general information about blocklabels. The data needed for each blocklabel are described in the section above.

The MODPATH-OBS blocklabels are listed in table A1. Input blocks need to occur in the order shown in table A1. Some may be absent, but those present need to occur in the order shown.

If a blocklabel is misspelled, the data are ignored and defaults assigned. Ignoring unneeded input blocks allows great flexibility for the sequences of runs common with MODPATH-OBS because most input blocks do not need to be removed even if they are not needed in a subsequent step. More generally, this feature allows different applications of the MODPATH-OBS to use the same or very similar input files. The drawback is that an input block is ignored if the blocklabel is misspelled.

Blockformat

The variable blockformat defines the structure of the data presented. The options are listed in table A2. The default blockformat is Keywords, but it is urged that the blockformat be listed specifically to reduce confusion.

The input blocks used in MODPATH-OBS are very flexible. One resulting difficulty is that if the blockformat specified does not match the format used, the information in the data block is ignored and generally no error message is printed. For example, if blockformat 'Keywords' is specified by default or designation, data organized in blockformat 'Table' is ignored. The problem can be detected by inspecting the echo of the input in the main MODPATH-OBS output file.

Table A2. Blockformat options.

Blockformat	Prescribed input format
KEYWORDS	Blockbody consists of a series of lines of the form: Keyword=value Under some circumstances there are restrictions on how the lines are ordered; see the input block instructions. If no blockformat is specified, KEYWORDS is assumed, but it is advisable to explicitly identify the block format to reduce errors. Comments are allowed.[1,2]
TABLE	Blockbody consists of a table of data that may have labels on the columns and may be read from the main input file or from another input file. See the text for additional information. Comments are allowed right after the BEGIN statement but not in the rest of the input block.[1]
FILES	Blockbody consists of the pathname for one or more files. Comments are allowed.[1,2] To allow the format to be specified, the contents of each of the listed files needs to begin with a 'Begin Blocklabel [Blockformat]' line and end with an 'End Blocklabel' line. The Blocklabel needs to be the same as in the 'Begin Blocklabel FILES' block within which the files are listed.

[1]Comments are separate lines starting with a # in the first column. No blank lines are allowed within any input blocks.

[2]Comments can be inserted anywhere within the input block.

Blockbody

blockbody contains data or the names of files from which the data are to be read. The format of the data is determined by *blockformat*.

The meaning of the data provided is defined using keywords. Keywords that are not recognized are ignored. This allows a constructed input block to be used for multiple purposes without modification. It also means that misspelled keywords are not flagged as errors and default values will be used if keywords are misspelled. This problem can be identified by reviewing the echo of the input file in the main MODPATH-OBS output file. For many keywords, a default is available and is used if the keyword is omitted.

Blockformat KEYWORDS

If *blockformat* is specified as KEYWORDS, blockbody is expected to be a series of phrases of the form keyword=value. For example, PARAMNAME=K1. There can be spaces on each side of the equal sign. Phrases can occur on separate lines or can occur on the same line if they are separated by spaces.

Some keywords can appear in any order while other keywords indicate the need for associated data to be provided either through a subsequent set of keywords or by other means. The options available depend on the input block, as described in the following chapters.

An example of a keyword that indicates the need for associated data occurs for blocklabel Options. Each time the keyword Verbose appears, a parameter is defined and a related set of data is needed. It can be tedious when a keyword and associated data are repeated for each parameter; blockformat TABLE is often more convenient in this circumstance (table A3).

Here is a simple example input block using blockformat keywords.

```
BEGIN Options Keywords
Verbose=0
Derivatives_Interface = "tcl.derint"
END Options
```

Blockformat TABLE

If *blockformat* is specified as TABLE, the first non-comment line of blockbody is in the format:

NROW=*nr* NCOL=*nc* [COLUMNLABELS] [DATAFILES=*nfiles*] [GROUPNAME=*gpname*]

The format of the rest of the blockbody depends on whether DATAFILES is listed, as shown in table A3.

Table A3. For blockformat *TABLE,* the format of blockbody after the first line without and with the optional keyword DATAFILES.

Without DATAFILES keyword			With DATAFILES keyword	
[column-name] [column-name]...		val	[column-name] [column-name]...	
val	... val	val	pathname	[SKIP=nskip]
...	... ...	...	pathname	[SKIP=nskip]
			...	
number of lines: nr			number of lines: nfiles	

Definition of keywords and variables:

NROW and NCOL are required keywords.

nr is the number of rows in the table.

nc is the number of columns in the table.

COLUMNLABELS is an optional keyword.

> COLUMNLABELS omitted: A default column order is used to identify the data in the columns of the table. Default column orders are only available for the *blocklabels* identified in section 'Blocklabels'. If a default column order is not available COLUMNLABELS is required.

> COLUMNLABELS listed: Column names are used to identify the data in the columns of the table. Data is read for columns with column names that are equivalent to keywords for this *blocklabel*. The keywords for each input block are defined in the following chapters. Data in columns with other labels are ignored. This allows data sets to contain columns that are not used by MODPATH-OBS. However, it also means that misspelled keywords are not flagged as errors and default values will be used if keywords are misspelled.

DATAFILES is an optional keyword.

> DATAFILES omitted: *nr* rows of data are read. Each *val* is a data value. The data type expected for *val* depends on the *blocklabel* and possibly on *column-name*. All data values for a row need to be on one line of the file. One line can contain up to 2,000 characters.

> DATAFILES listed: A list of file pathnames is read. The number of pathnames read equals **nfiles**, for example, DATAFILES=2. Each *pathname* is the path to a file from which rows of data are read. Paths with spaces need to be enclosed in double quotes. Each file needs to contain rows of data in columns in either the default column order or the order defined by the *column-name* entries, if specified. Data read from all files are combined as if read from one file. Each file is read in order until nr rows of data have been read. If *SKIP=nskip* is specified, *nskip* lines at the beginning of the file are ignored, and reading of data starts on the following line.

GROUPNAME is an optional keyword.

> For blocks that use groups, GROUPNAME=*gpname* can be used to assign a group name to all rows in the table. *gpname* is the group name. If GROUPNAME=*gpname* is present, GROUPNAME will not be in the default list of columns and cannot be included with the COLUMNLABELS option.

Here is a simple example input block using blockformat table.

```
BEGIN Parameter_Values TABLE
# These values override values in Parameter_Data input block
   nrow=9  ncol=2  columnlabels
   paramname   startvalue
   Wells_TR    -1.1000
   RCH_Zone_1  6.3072E+1
   RCH_Zone_2  3.1536E+1
   Rivers      1.2000E-3
   SS_1        1.3000E-3
   HK_1        3.0000E-4
   Vert_K_CB   1.0000E-7
   SS_2        2.0000E-4
   HK_2        4.0000E-5
END Parameter_Values
```

blockformat FILES

If blockformat is specified as FILES, the input block can contain one or more lines, each containing a pathname to a file. Lines with # as the first character are interpreted as comments and are ignored. Data read from all files in the list are combined to create one blockbody. The data need to be composed of blocks with Begin and End statements.

Data can be read from files in two ways. The mechanisms and their characteristics are described in table A4.

Table A4.   Alternatives for reading data from files.

Blockformat table With DATAFILES	Blockformat files
There is only one Begin blockformat and End blockformat block.	There can be more than one Begin blockformat and End blockformat block.
All data are read as a table.	Blockformat can change based on the designations in the Begin statements

Here is a simple example input block using blockformat files.

```
BEGIN OBSERVATION_DATA FILES
tc1.hed
tc1.flo
END OBSERVATION_DATA
```

Files tc1.hed and tc1.flo are read. For example, file tc1.flo might be as follows.

```
BEGIN OBSERVATION_DATA TABLE
  NROW=3  NCOL=4  COLUMNLABELS
  Obsname      obsvalue  statistic      equation
  flow.ss      -4.4      0.4            _
  flow.t3      -4.1      0.38           _
  flow.t12     -2.2      0.21           _
  flow.t3_ss    0.3      0.55           flow.t3 - flow.ss
  flow.t12_ss   2.2      0.45           flow.t12 - flow.ss
END OBSERVATION_DATA
```

Appendix B. Selected Input and Output Files for Hypothetical Example

For the transient-state example model the following main input (mpobs.in) and observation input data (obsdata.txt) are listed below. Additional explanatory comment lines were added to the main input file. If an option is not used, the data for that option is omitted by the user. If no child models are used the parent model data are the only input required.

MODPATH-OBS main input file (MPObsInput\mpobs.in)

```
Begin Options Keywords
  Verbose=0
  Run_Type=UCODE
  Time_Units=Years
  Reference_Time=2010
  Modpath_Control_File=mp.in
  Volume_Column=24
  TimeThresh=0.1
End Options

#########################################
########  Modpath Name Files  #########
#########################################
Begin Models Table
  Nrow=3 ncol=2 ColumnLabels
  Modpath_Namefile      Model_ID
  Parent\MODPATH_PARENT.dat   Parent
  Child1\MODPATH_CHILD1.dat   Child1
  Child2\MODPATH_CHILD2.dat   Child2
END Models

########### Parent Model ##############
#Define the river, trench, and farm
#open/close is the only option for the array data
Begin Source_Zones_Parent Table
  Nrow=1  ncol=3 ColumnLabels
  KTop KBot Array_Data
  1      1  "Open/Close .\parent\parent_source_array_lay1.dat 1 (FREE) -9"
End Source_Zones_Parent

Begin Observation_Zones_Parent Table
  Nrow=4 ncol=5 ColumnLabels
  Zone_ID JCol    IRow   KTop  KBot
     2      3       3      1     1
     3      3      19      1     1
     4      3      34      1     1
     5      3      45      1     1
End Observation_Zones_Parent
```

```
###########Child Model 1###############
Begin Source_Zones_Child1 Table
#Define the pit
  Nrow=1  ncol=3 ColumnLabels
  KTop  KBot Array_Data
   1     1   "Open/Close .\child1\child1.srz 1 (FREE) -9"
End Source_Zones_Child1


# Define Observation at Well Number 2 in Child model 1
Begin Observation_Zones_Child1 Table
  Nrow=1 ncol=5 ColumnLabels
  Zone_ID  JCol   IRow   KTop   KBot
      6    73     55      1      6
End Observation_Zones_Child1


###########Child Model 2###############
###Source_ Zones_Model_Child2 input block not needed, so omit.
#Define Observation at Well Number 3 in Child model 2
Begin Observation_Zones_Child2 Table
  Nrow=1 ncol=5 ColumnLabels
  Zone_ID JCol  IRow  KTop    KBot
   7      73    55    1       6
End Observation_Zones_Child2


########################################
#############Sources#################
########################################
# Source_Point input block not needed, so omit
Begin Source_Poly Table
#Face_No 6 is the top of the cell (fig. 1b).
  Nrow=5  ncol=3 ColumnLabels
  Loc_ID   Zone_ID   FACE_NO
  pit        1         6
  trench     3         6
  river      4         6
  farm       5         6
  farm2      2         6
End Source_Poly


Begin Observation_Point Table
Nrow=1  ncol=10 ColumnLabels
Loc_ID    MODEL_ID JCOL IROW KTOP  INDEX2 LOCX LOCY LOCZT LOC2
well1_obs parent     54   40   1      2   0.1  0.4  0.8   0.1
End Observation_Point


Begin Observation_Poly Table
#Face_No 0 is any face on the cell (fig. 1b)
  Nrow=6 ncol=3 ColumnLabels
  Loc_ID     Zone_ID     FACE_NO
  well2_obs     6           0
  well3_obs     7           0
  rivrch1_ob    2           0
  rivrch2_ob    3           0
  rivrch3_ob    4           0
  rivrch4_ob    5           0
End Observation_Poly
```

```
BEGIN OBSERVATION_DATA Files
Ucodeinput\obsdata.txt
END OBSERVATION_DATA

##########################################
#####     Concentration Files      ####
##########################################
Begin Concentration_Files Table
  Nrow=3 ncol=1 ColumnLabels
  FileName
  MPObsInput\cfc.dat
  MPObsInput\pcepit1.dat
  MPObsInput\pcetrench1.dat
End Concentration_Files

##########################################
#####     Exceedance Info          ####
##########################################
Begin Exceedance_Definiton Table
  Nrow=2 ncol=6 ColumnLabels
  ExcDefID StartTime EndTime Threshold NObs Operator
   Exc1      1970     2000       5        31   gt
   Exc2      2001     2010       5        10   gt
End Exceedance_Definiton

##########################################
#####     Output Files      ##########
##########################################
Begin Output_Files Table
  Nrow=8 ncol=2 ColumnLabels
  FType      FileName
  XYZDAT     MPObsOutput\xyz.dat
  TIMDAT     MPObsOutput\tim.dat
  CONDAT     MPObsOutput\conc.dat
  SRCDAT     MPObsOutput\src.dat
  XYZINS     UcodeInput\XYZUCODE.ins
  TIMINS     UcodeInput\TIMUCODE.ins
  CONINS     UcodeInput\CONUCODE.ins
  SRCINS     UcodeInput\SRCUCODE.ins
End Output_Files
```

MODPATH-OBS Observation data input file (Ucodeinput\obsdata.txt)

The file can be found in subdirectory UcodeInput, in file obsdata.dat. Here, the input for 36 of the 109 observations is listed.

```
BEGIN OBSERVATION_DATA TABLE
# This file is used both by MPATH_OBS and UCODE
# MPATH_OBS keywords: OBSType, Comp_Type, Obs10C_ID, SampleTime, Source_ID, NoPartValue
# UCODE keywords: GroupName, Statistic, Transform (for MPATH_OBS there is no group input block to
#               define Transform)
# Shared keywords: ObsName, ObsValue
# Weights generally 10% of obs with a minimum of 1.0. For proximity, multiply distance
#               between source and obs by 10%.
  Nrow=36 ncol=10 ColumnLabels
```

ObsName	OBSType	Comp_Type	Obs10C_ID	SampleTime	Source_ID	Transform	GroupName	ObsValue	Statistic
xyzobs1	Proximity	Xmin	well1_obs	2010	farm2	no	xyzobs	0	30
timmedpit2	Time	Med	well2_obs	2010	pit	no	timobs	16.4444	1.64
timge1003	Time	PctGe100	well3_obs	2010	all	no	timobs	2.54612	1
timlt1003	Time	PctLT100	well3_obs	2010	all	no	timobs	97.4539	9.74
cnc_cfc1	Conc	Conc	well1_obs	2010	cfc	no	cncobs	540.339	54.0
cnc_pce1	Conc	Conc	well1_obs	2010	pce	no	cncobs	0.00000	1
cnc_cfc2a	Conc	Conc	well2_obs	2010	cfc	no	cncobs	406.720	40.6
cnc_cfc2b	Conc	Conc	well2_obs	2009	cfc	no	cncobs	398.642	39.8
cnc_cfc2c	Conc	Conc	well2_obs	2008	cfc	no	cncobs	390.335	30.0
w2pce1970	Conc	Conc	well2_obs	1970	pce	no	cncobs	0.00000	1
w2pce1971	Conc	Conc	well2_obs	1971	pce	no	cncobs	4.39637	1
w2pce2009	Conc	Conc	well2_obs	2009	pce	no	cncobs	6.30325	1
w2pce2010	Conc	Conc	well2_obs	2010	pce	no	cncobs	5.69801	1
w2pceExc1	Conc	Exc1	well2_obs	1	pce	no	cncobs	93.5484	6.8
w2pceExc2	Conc	Exc2	well2_obs	2	pce	no	cncobs	100.000	5
cnc_cfc3	Conc	Conc	well3_obs	2010	cfc	no	cncobs	471.434	47.1
cnc_pce3	Conc	Conc	well3_obs	2010	pce	no	cncobs	0.646352	1
typfarmto1	Source	Percent	well1_obs	2010	farm	no	typobs	71.1176	7.1
typ_farm2	Source	Percent	well2_obs	2000	farm	no	typobs	68.2965	4
typ_river2	Source	Percent	well2_obs	2000	river	no	typobs	5.10194	1.1
typ_pit3	Source	Percent	well3_obs	2010	pit	no	typobs	0.00000	1
typtrench3	Source	Percent	well3_obs	2010	trench	no	typobs	0.893543	1
typ_farm3	Source	Percent	well3_obs	2010	farm	no	typobs	94.8501	9.2
typ_river3	Source	Percent	well3_obs	2010	river	no	typobs	1.54689	1
tyW2Pit2010	Source	Percent	well2_obs	2010	pit	no	typobs	2.06612	1
tyW2Pit1975	Source	Percent	well2_obs	1975	pit	no	typobs	9.39221	1
tyW2Pit1970	Source	Percent	well2_obs	1970	pit	no	typobs	0.00000	1
tyW2Tr2010	Source	Percent	well2_obs	2010	trench	no	typobs	11.0321	1
tyW2Tr2005	Source	Percent	well2_obs	2005	trench	no	typobs	9.11974	1
tyW2Tr2000	Source	Percent	well2_obs	2000	trench	no	typobs	6.01991	1
tyW2Tr1995	Source	Percent	well2_obs	1995	trench	no	typobs	6.50304	1
tyW2Tr1990	Source	Percent	well2_obs	1990	trench	no	typobs	7.10214	1
tyW2Tr1985	Source	Percent	well2_obs	1985	trench	no	typobs	5.46913	1
tyW2Tr1980	Source	Percent	well2_obs	1980	trench	no	typobs	3.16939	1
tyW2Tr1975	Source	Percent	well2_obs	1975	trench	no	typobs	0.00000	1
tyW2Tr1970	Source	Percent	well2_obs	1970	trench	no	typobs	0.00000	1

```
END OBSERVATION_DATA TABLE
```

Selected Partial MODPATH-OBS output files

The MODPATH-OBS Observation and Instruction output files for the transient-state hypothetical model as follows. Spaces have been omitted in some lines and significant figures of the numbers have been truncated to improve presentation and only some lines of output are shown here for illustration of output. The heading for each data set is the output filename. The listed quantities are the observation name, the simulated value (column 2), the observed value (column 3), and the residual (observed minus simulated; column 4), calculated as the observed minus simulated values.

XYZ.DAT

XYZOBS1	0.000	0.000	0.000

TIM.DAT

TIMMEDPIT2	16.163	16.444	-0.280
TIMGE1003	1.5248	2.5461	-1.021
TIMLT1003	98.475	97.453	1.021

CONC.DAT

CNC_CFC1	538.70	540.33	-1.637
CNC_PCE1	0.0000	0.0000	0.000
CNC_CFC2A	334.44	406.72	-72.27
CNC_CFC2B	328.29	398.64	-70.34
CNC_CFC2C	322.33	390.33	-67.99
W2PCE1970	0.0000	0.0000	0.000
W2PCE1971	3.0466	4.3963	-1.349
W2PCE2009	11.133	6.3032	4.829
W2PCE2010	10.017	5.6980	4.319
W2PCEEXC1	93.548	93.548	-0.15E-04
W2PCEEXC2	100.00	100.00	0.000
CNC_CFC3	504.97	471.43	33.53
CNC_PCE3	0.2316	0.6463	-0.4146

SRC.DAT

TYPFARMTO1	54.464	71.117	-16.653
TYP_FARM2	75.393	68.296	7.0966
TYP_RIVER2	3.7002	5.1019	-1.4016
TYP_PIT3	0.0000	0.0000	0.0000
TYPTRENCH3	0.3210	0.8935	-0.5725
TYP_FARM3	98.234	94.850	3.3842
TYP_RIVER3	1.4446	1.5468	-0.1022
TYW2PIT2010	2.7752	2.0661	0.7090
TYW2PIT1975	10.268	9.3922	0.8760

XYZUCODE.ins (Initially created by a run of MODPATH-OBS done outside UCODE)

```
jif @
11   [XYZOBS1]23:53
```

TIMUCODE.ins (Initially created by user and then updated by MODPATH-OBS)

```
jif @
11   [TIMMEDPIT2]23:53
11   [TIMGE1003]23:53
11   [TIMLT1003]23:53
```

CONUCODE.ins (Initially created by user and then updated by MODPATH-OBS)

```
jif @
11   [CNC_CFC1]23:43
11   [CNC_PCE1]23:43
11   [CNC_CFC2]23:43
11   [CNC_PCE2]23:43
11   [CNC_CFC3]23:43
11   [CNC_PCE3]23:43
```

SRCUCODE.ins (Initially created by user and then updated by MODPATH-OBS)

```
jif @
11   [TYPFARMT01]23:53
11   [TYP_FARM2]23:53
11   [TYP_RIVER2]23:53
11   [TYP_PIT3]23:53
11   [TYPTRENCH3]23:53
11   [TYP_FARM3]23:53
11   [TYP_RIVER3]23:53
11   [TYW2PIT2010]23:53
11   [TYW2PIT1975]23:53
11   [TYW2PIT1970]23:53
11   [TYW2TR2010]23:53
11   [TYW2TR2005]23:53
11   [TYW2TR2000]23:53
11   [TYW2TR1995]23:53
11   [TYW2TR1990]23:53
11   [TYW2TR1985]23:53
11   [TYW2TR1980]23:53
11   [TYW2TR1975]23:53
11   [TYW2TR1970]23:53
```

UCODE main input file (example_transient_ucode_reg.in)

This file is for the UCODE Parameter-Estimation Mode. Some comments and unused keywords have been removed from the printed file to improve the printed appearance. The unused keywords are included in the distributed file to make it easier to activate other options available in UCODE. A comparable PEST input file can be constructed.

```
# ------------------------
# UCODE INPUT EXAMPLE 1
# ------------------------
BEGIN Options
  Verbose=0
END Options


# ------------------------
# REGRESSION-CONTROL INFORMATION
# ------------------------
BEGIN UCODE_CONTROL_data KEYWORDS
  ModelName=example_transient
  MODELLENGTH = Meters
  MODELTIME = Seconds
#
  sensitivities=no            # Calculate sensitivities: yes, no
  optimize=yes                # Perform parameter estimation: yes, no
#
#Print sensitivities: css,dss(includes css),onepercentss,allss,unscaled,all,none
  STARTsens=dss               # For starting parameter values
  IntermedSENS=css            # For each parameter estimation iteration
  FinalPrint=dss              # For final parameter values
#
#Print residuals: yes/no
  StartRes=yes                # Starting parameter values
  IntermedRes=yes             # For each parameter estimation iteration
  FinalRes=yes                # Final parameter values
#
  DataExchange=yes            # Graphing & postprocessing: yes, no
END UCODE_CONTROL_data

BEGIN REG_GN_CONTROLS KEYWORDS
# How to end GN iterations. (SOSWR=sum-of-squared weighted residuals)
  tolpar=0.07                 # frac parameter value change for convergence
  tolsosc=0.01                # frac SOSWR change over 3 GN iters for converge
  maxiter=100                 # maximum # of GN iterations
#
#Other keywords determine performance during GN iterations
  maxchange=0.3               # max frac parameter change between iterations
  maxchangerealm=regression   # Values for maxchange & tolpar: Native,Regression
# Marquardt parameter
  MrqtDirection=85.411137668  # angle (in degrees) above which Mrqt par used
  MrqtFactor=1.5              # a in newm=a(oldm)+b
  MrqtIncrement=0.001         # b in newm=a(oldm)+b
# QuasiNewton updating
  quasinewton=no              # Option to use quasi-newton updating: yes, no
    qniter=5                  #  # iterations before starting QN updating.
    qnsosr=0.01               # Frac SOSWR change over 2 GN iters to start QN
```

```
# Trustregion updating
  TrustRegion=hookstep           # Option to use truxt region: Dogleg/Hookstep/No
    MAXSTEP=1.e-3                 # Maximum step size
    consecmax=5                  #
# Conditions for omitting parameters
  OmitDefault=1                  # # of values to read. Omit obs with these values
  OmitInsensitive=yes            # yes: Omit insensitive pars. Check each iter.
    MinimumSensRatio=2.E-2       # Omit pars if CSS<MinimumSensRatio x CSSmax.
    ReincludeSensRatio=5.E-2     # Return pars if CSS>ReincludeSensRatio x CSSmax
END REG_GN_CONTROLS

BEGIN MODEL_COMMAND_LINES
# Single quotes around 'Command=value' required if the
# command includes spaces, optinal otherwise
  'Command= doit.bat'
  purpose=forward
  CommandId=mflgr-mpath
END MODEL_COMMAND_LINES


# --------------------
# PARAMETER INFORMATION
# --------------------

BEGIN PARAMETER_GROUPS KEYWORDS
    GroupName = hyd_prop  transform=yes
    GroupName = mlt_prop  transform=no
END PARAMETER_GROUPS

BEGIN PARAMETER_DATA FILES
  .\parameter\param_hyd.txt
  .\parameter\param_mlt.txt
END PARAMETER_DATA


# --------------------
# OBSERVATIONS
# --------------------

BEGIN OBSERVATION_GROUPS TABLE
###Input block for UCODE and MODPATH-OBS. Unrecognized keywords ignored.
###    CovMatrix used only by UCODE.
###    Wtcorrelated and printeach used only by MODPATH-OBS.
###Statistic and statflag defined here are not used for headobs
###Plotsymbol defined for colors in gw_chart for dss (_sd) and dfbetas (_rb)
###Report version excludes WtMultiplier and Transform
  nrow=5 ncol=7  columnlabels
groupname statistic statflag plotsymbol useflag  WTCORRELATED PRINTEACH
  headobs    1.0        sd        16         yes        no          no
  xyzobs     4.47       sd        14         yes        no          no
  timobs     4.47       sd        1          yes        no          no
  cncobs     4.47       sd        2          yes        no          no
  typobs     4.47       sd        12         yes        no          no
END OBSERVATION_GROUPS

BEGIN OBSERVATION_DATA FILES
    UcodeInput\head.obs
    UcodeInput\obsdata.txt
END OBSERVATION_DATA
```

```
# ---------------------
# PRIOR INFORMATION
# ---------------------

BEGIN PRIOR_INFORMATION_GROUPS TABLE
nrow=2 ncol=6 columnlabels
  groupname   statistic statflag   plotsymbol useflag  wtmultiplier
  T             0.5        SD            6       yes        1.0
  P             0.13       SD            7       yes        1.0
END PRIOR_INFORMATION_GROUPS

BEGIN LINEAR_PRIOR_INFORMATION TABLE
nrow=10  ncol=4 columnlabels
PriorName     Equation        PriorInfoValue  GroupName
T110Prior     log10(T110)     1.0E-07              T
T70Prior      log10(T70)      5.0E-07              T
T30Prior      log10(T30)      6.0E-06              T
T16Prior      log10(T16)      5.0E-05              T
T08Prior      log10(T08)      1.0E-04              T
P110Prior     log10(P110)     0.20                 P
P70Prior      log10(P70)      0.25                 P
P30Prior      log10(P30)      0.25                 P
P16Prior      log10(P16)      0.25                 P
P08Prior      log10(P08)      0.25                 P
END LINEAR_PRIOR_INFORMATION


# ---------------------
# MODEL FILES
# ---------------------

BEGIN MODEL_INPUT_FILES TABLE
  nrow=9 ncol=2 columnlabels
   modinfile                  templatefile
   .\parent\parent.por        .\template\parent_por.tpl
   .\child1\child1.por        .\template\child1_por.tpl
   .\child2\child2.por        .\template\child2_por.tpl
   .\parent\Parent_MULTp.mlt  .\template\Parent_MULTp.tpl
   .\child1\Child1_MULTp.mlt  .\template\Child1_MULTp.tpl
   .\child2\Child2_MULTp.mlt  .\template\Child2_MULTp.tpl
   .\parent\PARENT.pvl        .\template\PARENT_pvl.tpl
   .\child1\CHILD1_tran.pvl   .\template\CHILD1_tran_pvl.tpl
   .\child2\CHILD2_tran.pvl   .\template\CHILD2_tran_pvl.tpl
END MODEL_INPUT_FILES

BEGIN MODEL_OUTPUT_FILES TABLE
  nrow=7 ncol=3 columnlabels
   modoutfile                 instructionfile          category
   .\MPObsOutput\XYZ.dat      .\UCodeInput\XYZUCODE.ins    obs
   .\MPObsOutput\TIM.dat      .\UCodeInput\TIMUCODE.ins    obs
   .\MPObsOutput\CONC.dat     .\UCodeInput\CONUCODE.ins    obs
   .\MPObsOutput\SRC.dat      .\UCodeInput\SRCUCODE.ins    obs
   .\parent\pheadobs.txt      .\UCodeInput\PHead.ins       obs
   .\child1\c1headobs.txt     .\UCodeInput\C1Head.ins      obs
   .\child2\c2headobs.txt     .\UCodeInput\C2Head.ins      obs
END MODEL_OUTPUT_FILES
```

```
# -------------------------------
# PARALLEL-PROCESSING INFORMATION
# -------------------------------

BEGIN PARALLEL_CONTROL
  PARALLEL=yes            (default=no)
  WAIT=0.500              (default=0.001)
  VERBOSERUNNER=5         (default=3)
  AUTOSTOPRUNNERS=yes     (default=true)
  OPERATINGSYSTEM=Windows
  TIMEOUTFACTOR = 2.0
END PARALLEL_CONTROL

BEGIN PARALLEL_RUNNERS TABLE
# RUNNERDIR must end with the correct directory separator for
#   the OS -- "\" for Windows and "/" for Unix and Linux.
#
# Table could contain columns COPYMODELIN and COPYMODELOUT -- NOT IMPLEMENTED
#   to indicate ftp or copy command scripts to be run to copy
#   model input before model execution, and model output after
#   model execution.  This, when supported, will permit ftp to
#   be used, and so add support for parallel runs to involve
#   Unix or Linux machines.
#
# The pathnames are relative to the batch file running UCODE, not the location
#   of this input file.
#
  NROW=5  NCOL=3  COLUMNLABELS
  RUNNERNAME   RUNNERDIR                            RUNTIME
  runner1      .\runners\runner1\Example\           20000
  runner2      .\runners\runner2\Example\           20000
  runner3      .\runners\runner3\Example\           20000
  runner4      .\runners\runner4\Example\           20000
  runner5      .\runners\runner5\Example\           20000
END PARALLEL_RUNNERS
```

UCODE PARAMETER_DATA input block (in files (1) param_hyd.txt and (2) param_mlt.txt)

(1) param_hyd.txt (Transmissivity parameters)

```
BEGIN PARAMETER_DATA TABLE
# If constrain=yes, need lowerconstranint and upperconstraint
#      1          2          3          4        5         6          7          8
  nrow=5  ncol=8 columnlabels  GroupName=hyd_prop
  paramname  startvalue  lowervalue uppervalue  constrain adjustable perturbamt transform
  T110       1.0E-07    1.0E-8     1.0E-6        no        yes        0.02       yes
  T70        5.0E-07    5.0E-8     5.0E-6        no        yes        0.02       yes
  T30        6.0E-06    6.0E-7     6.0E-5        no        yes        0.02       yes
  T16        5.0E-05    5.0E-6     5.0E-4        no        yes        0.02       yes
  T08        1.0E-04    1.0E-5     1.0E-3        no        yes        0.02       yes
END PARAMETER_DATA TABLE
```

(2) param_mlt.txt (Porosity and Specific-Storage parameters)

```
BEGIN PARAMETER_DATA TABLE
#If constrain=yes is used, need lowerconstraint and upperconstraint
#      1          2          3          4        5         6          7          8
  nrow=6  ncol=8 columnlabels  GroupName=mlt_prop
  paramname  startvalue  lowervalue uppervalue  constrain adjustable perturbamt transform
P110         0.20       0.11       0.36          no        yes        0.02       yes
P70          0.25       0.14       0.45          no        yes        0.02       yes
P30          0.25       0.14       0.45          no        yes        0.02       yes
P16          0.25       0.14       0.45          no        yes        0.02       yes
P08          0.25       0.14       0.45          no        yes        0.02       yes
SSKEC        5.0E-06    1.0E-07    1.0E+02       no        no         0.25       no
END PARAMETER_DATA TABLE
```

Appendix C. Program Distribution and Installation

Appendix C describes the distributed files and directories, how to compile and link the source code to obtain an executable MODPATH-OBS file, issues of portability, and memory requirements for MODPATH-OBS.

Distributed Files and Directories

MODPATH-OBS is used with MODPATH-LGR, MODFLOW-LGR, and MODPATH (version 4.0 or later), and can be downloaded from the web site listed in the preface. The operating system is listed for each compiled downloadable executable file. When uncompressed, a directory is created with four subdirectories. The subdirectories are listed in table C1.

Table C1.   Contents of the subdirectories distributed with MODPATH-OBS.

[Name files, files used by MODFLOW-2000 to define program performance and input files.]

Sub-directory	Contents
bin	Executable file for MODPATH-OBS. Executables for MODFLOW and UCODE are in the bin directory under test-data-win.
doc	This documentation file, in PDF format.
src	Fortran source files for MODPATH-OBS. Most source files are named with the extension "f". Source files from the JUPITER API have extension "f90".
test-data-*win*	Subdirectories contain all files and executables to run a set of examples problems. The examples include the steady state and transient flow field versions of the hypothetical example problem described in the text. After the examples are run using the batch files provided, these subdirectories also include process-model output files.

Subdirectory	Contents
bin	Executables for all programs needed to run the example problems.
Example_SS	Files for the example using a steady-state flow field.
Example_Transient	Files for the example using a steady-state flow field

The above two directories both have the following subdirectories	
child1	MODFLOW and MODPATH files for child grid 1.
child2	MODFLOW and MODPATH files for child grid 1.
MPObsInput	Input files for MODPATH-OBS
MPObsOutput	Output files for MODPATH-OBS
parallel-sos	Directory used for UCODE run to investigate the objective function surface. Includes file sos-concat.bat to concatenate results from runs executed on different cores.
parameter	Files with UCODE Parameter_Data input blocks.
parent	MODFLOW and MODPATH files for the parent grid.
runners	Empty directory that can be populated using one of the ##-a-runners-populate.bat files, where ## is replaced by 01 for pre-regression sensitivity analysis, 02 for regression, or 03 for post-regression sensitivity analysis.
template	Template files that can be used by UCODE or PEST to create model input files with correct parameter value substitutions.
UcodeInput	UCODE main input files, instruction files, and files containing Observation_Data input blocks.
UcodeOutput	Directory to which UCODE output is directed. Contains file RunResidAnalysis.bat that uses UCODE output to run post-processor Residual_Analysis.
finegrid_ss	Steady-state flow problem using a uniformly file model grid.
finegrid_transient	Transient flow problem using a uniformly file model grid. Results from this run are used as observation.

Compiling and Linking

If changes to the source codes are needed, or if the codes are used with an operating system other than those for which executable files are distributed, the codes need to be compiled. The modules needed to compile each of the distributed codes are listed in a readme file located in the subdirectory for each code. These subdirectories are located within the src subdirectory.

The distributed source code is compatible with standard Fortran 90 and Fortran 95 except for the following:

1. The call to the SYSTEM subroutine, which is used to initiate execution of an operating-system command. This call is in subroutine UTL_SYSTEM.

2. The call to the GETCL subroutine, which provides access to the command line used to invoke a program. This call is in subroutine UTL_GETARG

These subroutines are non-standard and compiler-dependent. Both subroutines are in the utilities module (UTL.F90) of the JUPITER API. It is expected that any changes needed to accommodate compilers that have different subroutines or different syntax for these capabilities would be restricted to these subroutines.

The object files created during compilation need to be linked to create an executable program. The linker program commonly is invoked as part of the compilation procedure.

In directories Example_SS and Example_TR, batch files are provided to set up the directories needed for parallel processing. The transient flow-field example is rather lengthy and use of parallel processing is advised. The set up for parallel processing provided for these examples is expected to be useful for a wide range of problems.

Other batch file in the Example_SS and Example_TR directories are designed to conduct a standard set of model calibration runs. The batch files are listed in table C2.

An annotated version of the global endpoint file is provided in the file global_endpoint-annotated.dat. This can be useful in evaluating endpoint files.

The file mp.in is the file used by MODFLOW-LGR to integrate the parent and two child grids.

Table C2.   Batch files distributed in directories Example_SS and Example_Transient

Batch file suffix (extension is always .bat)	Description (files ending with .in are UCODE main input files from the Ucode Input subdirectory
00-a-clean	Removes output files
00-b-ucode_run_example-forward	Conducts a forward model run using example_ucode_forward.in
01-a-runners-populate	Populates directories used to run UCODE in parallel
01-b-ucode_clean&start_runners	Removes some output files and starts runners in each directory.
01-c-ucode_run_example_sensitivity-analysis	Executes UCODE to conduct pre-regression sensitivity analysis using parallel processing
02-a-runners-populate	Populates directories used to run UCODE in parallel
02-b-ucode_clean&start_runners	Removes some output files and starts runners in each directory.
02-c-ucode_run_example_regression	Executes UCODE to conduct regression using parallel processing
doit	Runs MODFLOW, MODPATH, and MODPATH-OBS
doit-DP	Same as doit using double precision executables.
doit-extra-printing	Same as doit with extra printing to evaluate MODPATH-OBS performance.
mpath_obs	Runs MODPATH-OBS
mplgr	Runs MODPATH-LGR
RunModflow.bat	Runs MODFLOW
Runners-01a	Used by ##-a-runners-populate.bat files to populate runners used for parallel UCODE runs.
RunResidAnalysis.bat	Runs the UCODE residual analysis post-processor

Portability

MODPATH-OBS, MODFLOW-LGR, and MODPATH-LGR were written in standard Fortran 90/95. A modular style is used to enhance accuracy, to simplify maintenance, and to encourage innovation. The aspects of the code mentioned above that are compiler dependent are also platform dependent.

Memory Requirements

As distributed, the source files and executable file dynamically allocate memory. Thus, the program automatically adapts to whatever memory is required and no user intervention is required. Slow execution times can result if the memory required exceeds the physical memory available on the computer being used. Both 32-bit and 64-bit versions are provided with the distribution of MODPATH-OBS.

Appendix D. Add Volume: A Utility Program to Assign Volume to Particles

AddVolume is a utility program which assigns volumes to particles. This supports the calculation of concentrations, among other things.

AddVolume adds a column to either the starting locations or endpoint file of MODPATH. This column lists a value for volume associated with the associated MODPATH particle. The volume is determined by AddVolume as follows.

- Read the cell-by-cell budget file produced by MODFLOW.

- Read the MODPATH input file and count the number of particles started on each cell face.

- Assign each particle a volume equal to the flow across the face divided by the number of particles started on the face.

AddVolume works best with particles that are started on the faces of model cells that have sources (for forward tracking) or sinks (for backward tracking).

The AddVolume program is run from the command line with two required arguments and one option argument, as follows. Here the optional argument is italicized.

AddVolume FileType MPNameFile *DoublePrecision SFilename*

Explanation of the Arguments

FileType—The two possible options are StartingLocations and Endpoint.
  StartingLocations: Add the volume to the starting locations file. This is specified in the response file or name file in Modpath-5 and previous or the simulation file in Modpath-6.
  Endpoint: Add the volume to the endpoint file. This is specified in the name file (or is endpoint by default) in Modpath-5 and previous or the simulation file in Modpath-6.
Tracking–The two possible options are Forward and Backward (see modpath name file).
  Forward: Add the volume to the MODPATH starting locations file.
  Backward: Add the volume to the MODPATH endpoint file.

MPNameFile–The MODPATH Name File

DoublePrecision–This is an optional item that needs to be specified when the MODFLOW cell-by-cell budget file was saved using double precision. True if the cell-by-cell budget file from MODFLOW was saved using double precision, false for single precision. Default=False.

SFileName–For MODPATH 6, this is the name of the simulation file and is required. For earlier versions, this is an optional item that needs to be specified when FileType = StartingLocations and the Locations keyword does not appear in the Modpath name file. If needed in this case, SFileName is the filename of the file in which the starting locatings are listed.

Appendix E. ModpathParameters: A Utility Program to Assign Porosity Parameters Using Zone and Multiplier Arrays from MODFLOW

ModpathParameters is a utility program which allows a user to assign porosity parameters for Modpath in a similar way to which parameters are defined in MODFLOW. Parameters are defined based on arrays from the MODFLOW multiplier and zone arrays. The program uses the specified multiplier and zone arrays along with the parameter value to compute the porosity value for each cell. The final values are then written into the MODPATH main file.

If the MODFLOW model uses the HUF package, the parameters can be based on the hydrogeologic units defined there. For model cells that represent more than one hydrogeologic unit, the porosity is calculated as a thickness-weighted average of the values from each hydrogeologic unit.

This program will only run on a Windows operating system and requires the .NET Framework version 3.5

The ModpathParameters program is run from the command line with the name of the input file as the argument.

ModpathParameters Input File

The input file consists of the following items. The items each need to start on a new line in the input file.

1. MPNameFile

2. MFNameFile

3. UseHUFLayers

4. NP

Need NP sets of items 5 and 6

5. ParName ParType ParValue NClus

Need NClus sets of item 6 for each item 5.

6. LayerName MltArr Arr MODPATH name file.

Explanation of the Variables in the ModpathParameters Input File

MFNameFile–The name of the MODFLOW name file. This is used to get the Zone and Multiplier array packages and any data files they might depend on.

UseHUFLayers–set to true if the parameters are to be applied by hydrogeologic units defined in the HUF package; set to false to apply the zones based on the model layers.

NP–the number of parameters to be defined.

ParName–the name of the parameter.

ParType–the type of the parameter. This will be MPPOR for modpath porosity parameters.

ParValue–the value of the parameter.

NClus–the number of clusters used to define the parameter. Each line 6 record is a cluster (LayerName MltArr ZonArr IZ)

LayerName–the model layer (if UseHUFLayers = false) or the name of the hydrogeologic unit name (if UseHUFLayers = true)

MltArr–is the name of the multiplier array to be used to define array values that are associated with a parameter. The name "NONE" means that there is no multiplier array, and the array values will be set equal to Parval.

ZonArr–is the name of the zone array to be used to define array elements that are associated with a parameter. The name "ALL" means that there is no zone array and that all elements in the hydrogeologic unit or model layer are part of the parameter.

IZ–is up to 10 zone numbers (separated by spaces) that define the array elements that are associated with a parameter. The first zero or non-numeric value terminates the list. These values are not used if ZonArr is specified as "ALL".